Animal Husbandry

and

Other Social Graces

ISBN 0-9659279-0-3

Library of Congress Cataloging in Publication Data Pending

Lithographed and bound in the
United States of America by
The Caxton Printers Ltd.,
Caldwell, Idaho 83605

Contents

CHAPTER 1
SET A SPELL

Give a boy a gun

By Suzanne Huxhold

CLOVER CREEK-My husband and I have a running argument over guns. He likes them, I don't. I see them as dangerous, he sees them as necessary. I see them as killing machines, he sees them as tools, like a stapler, say, or a computer; if you used staplers and computers to beat things to death, that is.

But we did agree our son Joseph should learn to use a gun, eventually. Our lifestyle requires it. Even I can see how foolish it would be to live way out here in the wilderness, amongst bears and cougars and bulls who refuse to be moved, without a working knowledge of weapons. Even I could shoot a gun, if I had to. If I were, for instance, held at gunpoint.

We decided we'd get Joseph a BB gun when he was 12. Twelve seemed the perfect age to both of us.

So last year, when Joseph turned 7, Scott bought him a BB gun.

I fretted, but it seemed my husband was the wiser of us.

Joseph was responsible and serious with his little weapon of destruction. He learned to carry it properly, discharge it safely and treat it with respect. He also killed every box elder bug and cricket he could find. Just blew their little brains out. The exterminator, we called him. It was, apparently, very satisfying for him. I was surprised he didn't tie their tiny corpses to the hoods of his toy cars, frankly.

I still told them both every day or so that he'd shoot his eye out, but for the most part I got over being worried. He never popped the cats, never pointed his gun at the house, never held anyone for ransom.

Which is why I wasn't too concerned yesterday when I saw my little boy - BB gun in hand - and his pal J.J. crouched behind the stock tank in the heifer pen.

Until I watched a heifer walk up to get a drink and the two boys jumped up, silent as stealth bombers, and Joseph pegged that little heifer right in the head.

I stared out the window, dumbfounded. The heifer jolted a little, shook her head, and jogged off. The boys crouched down again, giggling like fools and shushing each other frantically.

They waited patiently until another heifer came to water. Just as they sprang up for the ambush, my husband, who had been watching from another window, roared his son's name.

Oh, the guilty look. Oh, the nervous giggling. Oh, the fear of retribution.

And that was mostly from Scott. I glared at him.

"See," I said, puffed up with righteous indignation. "I told you this would happen."

"He didn't shoot his eye out," my husband pointed out.

"He SHOT your heifer!" I told him, at the top of my lungs.

"It didn't hurt her."

"He SHOT your heifer!" I bellowed.

"You already said that."

"I'm just not sure you heard me."

"Honey, I heard you. See, my ears are bleeding."

We watched the boys trudge across the road toward the house.

I looked at my husband. He was waffling, as fathers will with mischievous sons. I watched his eyes soften. "He was just showing off in front of J.J.," he said. "He got carried away."

"He's a hoodlum. You know that, don't you? Our son is Lee Harvey Oswald. You know, if Lee Harvey Oswald actually shot Kennedy, which I personally don't believe, but you get my drift."

"We'll just take his gun away for a while."

"Six months," I said firmly.

"Until the weekend," my husband said at the same time.

I sputtered. "This is THURSDAY!"

The boys appeared at the back door. Scott furrowed his brow menacingly, took the gun, and ordered the boys upstairs to stew.

They were fooled. I was not. I know an indulgent parent when I see one.

"He SHOT your heifer!" I reminded him.

"Don't say that again," Scott warned.

We argued the sentence down to a month, with any further infractions meaning an all-out ban on weapons until the boy is 12. I'm firm on this point. Scott says maybe. Of course, he said, four years is a long time.

We'll be overrun by bugs by then.

Borrow me a roll of duct tape, please

By Steve Crump

I spent part of my childhood living in a trailer park, and I'm proud of it.

Trailer-park people are, by and large, honest, genuine and unassuming.

OK, they're *real* unassuming.

They don't put on airs, because they need them to pump up the bald tires on the '72 Impala sitting outside.

But some mobile home parks tend to have – now how to put this diplomatically? – certain quirks.

Blue-green Astroturf in the bathroom; posters of Tanya Harding and Paula Jones stapled to the family-room walls; counted cross-stitched samplers hanging in the kitchen that read, "Good bread, good meat, good gosh, let's eat!" – that sort of thing.

And the foremost trademark of trailerparkdom is duct tape.

To live in some mobile home preserves is to dwell among 2-inch-wide gunmetal-gray stripes. They hold everything together: '56 Ford pickups, leisure suits, windows, swamp coolers, Naugahyde recliners, faux-wood paneling – even marriages.

And while it's entirely possible to take a boy out of a trailer park, can you really take the trailer park out of the boy?

I mention this because when I recently noticed a rip in the downstairs sofa, my first instinct was to reach for the duct tape.

It seemed so natural, just to whang a couple of loops of Sheetmetal Worker's Grade No. 4 around that couch cushion and call it good enough for government work.

But then I was seized by horror, at just the moment that my wife seized the roll of tape from my trembling hand.

"Are you *mad*?" she gasped, wide-eyed.

For she, too, grew up in a mobile home park and understands that the slippery slope toward watching "Hee-Haw" reruns at 3 in the morning starts with just a smidge of duct tape.

It's so seductive. Will anyone really *notice* if there's duct tape wrapped around the hinges of the waffle iron?

Will they care that duct tape is all that's preventing the tail pipe from meeting the road? Can they honestly say that no bean-bag chair in *their* homes has ever lived beyond its years thanks to the timely intervention of Scotch-brand adhesives?

No, they cannot say those things, and no, they probably won't notice. But it doesn't matter.

You'll know.

Surrender to duct tape, and you're as good as admitting that you don't care if there's livestock in your living room, cobwebs on your dentures and rust on your razor.

To abuse duct tape is the moral equivalent of wearing sweat pants every day: It's abject surrender to the potbellied, beer-belching, evil angels of your nature.

For this much is certain: Those who duct-tape today are doomed to scratch unsocial places tomorrow.

So I'm here today to report proudly that the fissure in my sofa cushion remains untaped and widening.

Wonder if I could borrow some balin' wire?

Baseball fever hits tee-ball mom from hell

By Suzanne Huxhold

CLOVER CREEK — I'm sorry to see baseball season end in our little town. Who knew we had so many potential phenoms?

OK, maybe not so many. And maybe not even phenoms, but everyone had a good time, and that's what counts, right?

By everyone, I mean the spectators, by the way, not the ballplayers. Oh, I'm sure the ballplayers had a wonderful experience - good sportsmanship and fair play and healthy competition and all that - but it was the crowd that truly benefitted from the season.

It was my first encounter with baseball fever in Bliss. Until this summer my son was too young to play, and sitting in the blistering, splintering bleachers during the hottest part of the day, watching my neighbors' kids hit pop flies and spit Gatorade at each other in the dugout didn't strike me as a lot of fun.

But my fellow townspeople never lost an opportunity to tell me baseball season was the best part of the summer. I was skeptical. Even, I admit, a little disdainful.

I grew up with baseball. My family is a baseball family. My grandfather, the story goes, died of heart failure waving my uncle home from his position as third base coach. My brothers play, my cousins play, even my father plays, but that's mostly for the beer,

I think. You can justify a heck of a lot of beer drinking to your wife when you're playing ball.

It's old hat, baseball. I've been watching it all my life.

I was sitting next to my friend Carol at Joseph's first tee-ball practice, relating all this to her. She nodded politely.

"Uh-huh," she said. "So I'm sure you'll be bored with all this?" She gestured toward the noisy knot of 6-and-7-year-olds sitting cross-legged on the grass, trying to grasp the severity of the consequences of throwing the bat as described in gory detail by their patient coach.

"Oh, yes," I said, blandly, stifling a small yawn. "I'm just doing this because all Joseph's friends are here. I'm sure he'll be tired of it in a week or so, and we can forget all about it."

She smiled at me. "And it's all pretty unimportant, right?

Pretty silly?"

"Well, Carol, you must admit, it's not cancer research," I said. "It's baseball. It's not even baseball, it's tee-ball. The fact that everyone gets so wound up about it is ridiculous. That woman over there actually keeps STATS, for crying out loud. No offence, Carol, but these mothers here need to get themselves some lives."

She smiled again, her eyes drifting back to the tiny ballplayers, all earnestly engaged now in running around the bases, shouting, "FIRST, SECOND, THIRD, HOME" as they stomped on each base. I noticed Joseph skipped second and went straight to home.

"Joseph, honey, you skipped second," I shouted through cupped hands. "Go on back to second. There you go. No, now over to third.

That's right. GOOD JOB."

I turned back to talk to Carol, but she had her head in her hands, laughing.

"Suzi, you're doomed," she said. "You don't know it yet, but you're doomed."

I was doomed. Within a week I'd transformed myself from a reasonable, slightly bored spectator into TEAM MOM FROM HELL.

"Oh, come on, ump. Under no stretch of the imagination did that ball go ten feet from the tee," I was screaming by the second game. "Come on, you're letting her go? Oh, man. GET SOME GLASSES, YOU BUM."

By the third game, I was a little less inhibited.

"HEY UMP, CHECK THAT KID'S DRIVER'S LICENSE. IF HE'S 7, I'M 18, UMP, YOU BIG DUMMY," I shouted. "COME ON, COME ON, MOVE 'EM THROUGH. MY KID HASN'T EVEN GOT TO BAT YET. LET'S GO!"

By the fourth game, I was sitting in the dugout. The other mothers wouldn't let me sit in the bleachers with them anymore.

Wusses.

"Hey, Shawna," I leaned over to the coach's wife and peered at her stat sheet, "I'm pretty sure Joseph hit a double on that last at-bat."

"Uh, I'm sorry, but you don't count it as a double if there's an overthrow at first, Suzi," she answered politely.

"TIME," I stood up. "TIME, UMP. I WANT TO LODGE A FORMAL PROTEST OVER OUR STATISTICIAN. SHE, ummmphhh...," my husband clamped his hand over my mouth and dragged me to the car before I could finish. The other mothers made him do it, I'm sure.

Tee-ball only lasted five weeks, darn it. Best five weeks of my life, though, I must say. Next year I may just bring beer.

Portrait of guys standing around by their trucks

By Steve Crump

I finally found the picture they ought to put on the cover of the Idaho Travel Guide.

It's a photo of a bunch of guys standing around next to their trucks with their hands in their pockets.

Why, it's the portrait of the real West, as true to the spirit of Idaho as lyin' is to fishin'.

Standing by your rig with your hands in your Levi's is a proud tradition, a badge of honor that says, "I know where the four-point bucks are, and you're from California."

It's a gesture of great subtlety that's not, in my opinion, sufficiently appreciated by those Connecticut connoisseurs of Western folkways who commute to New York City in their Range Rovers and wear Tony Lamas with a three-piece suit.

And it stems from the well-known fact that Idaho men don't know what to do with their hands.

An Idahoan's hands are great inarticulate appendages good for workin' and swingin' long-necks, but worthless for conversation. Europeans, interior decorators, male hairdressers and politicians talk with their hands; real men hide them in their trousers.

It's a phenomenon, I think, that comes from the natural wariness of folks who grow up a long way from anyone else.

In a land of endless horizons, watching a trail of dust grow into a speck and then a figure and then a rider on a horse is a vaguely disquieting experience. Here comes a stranger, and before you extend your hand, you'd dang well better know what he's doing in your little corner of the High Lonesome.

But with women and civilization came the demand for social graces. If you went to church and hung around out by the outhouse, folks would think you were loco. So you had walk up to perfect strangers and make conversation.

("Hi you?")
("Don't do no good to complain.")
("How's your crop?")
("Done OK.")
("Been to the fair?")
("Already seen it.")

Still, strangers were not to be embraced — not yet. You have to size a fella up before you put him down as a friend.

So you'd stand there, hands anchored in your jeans, you and the stranger — like a pair of wooden Indians on a merry-go-round, slowly circling each other until one of you found your way to your horse and to safety.

The horses have gone to pasture, but old habits die hard. Fellas still do the cowboy minuet, casting sidelong glances at the truck that keeps their universe in perspective.

Hand management has evolved, however, thanks to belt loops and Western buckles.

Sometimes you just can't trust your hands to stay in your jeans, so you hook them on the outside — if you're really insecure, with your thumbs.

You can even tell where a man's from by the way he strikes the pose.

Eastern Idahoans, for example, used to nine months of standing in deep slush and long talks with the stake president, prefer the deep-pockets approach, giving them better leverage to shift from one foot to another. That way, you've always got feeling in one foot or another.

They stand a minimum of 10 yards from their Internationals, which they buy because of the four-wheel drive and the fact that it's only dealership in Ririe.

But Magic Valleyites are a different breed. Right hand hooked on a belt loop, left stuck tentatively into the front pocket, they kick little divots in the parking lot with the heel of their right boot, leaning against the cab of a battered GMC pickup with four balding tires.

The rig got that way from hauling lava rock.

At the end of the conversation, the Magic Valley man turns like a center pivot and eases away down the far left side of the road — the side away from the canal — moving like a shock absorber in search of a jolt.

At trail's end, there's Orion Samuelson, two dogs under the porch and french fries with brown gravy.

Shoot, a man got no reason to go to town anymore.

What a load of shirts

By Suzanne Huxhold

CLOVER CREEK-I was stuffing clothes into the washer the other day when my son came running into the laundry room.

"Mom," he yelled frantically, not because he was actually frantic, but because that's how he talks now that he's 7 and has figured out I'm too stupid to understand him unless he's loud and rude, "make sure you check all the pockets."

"It's just a load of shirts," I told him.

"Well, shirts have pockets."

I decided this kernel of obvious wisdom should be taken into account, as the last time I ignored something my child said about the wash, I ended up having to call a repairman to flush the remains of a baby bull snake out of the spin bucket.

I checked the pockets of the shirts.

In the pocket of Joseph's purple Hawaiian shirt I found a plain gray rock, three pennies and a AAA battery. In the pocket of his blue San Diego Zoo shirt I found a folded-up subscription insert for Field and Stream magazine with his name carefully printed in the address box, a plastic horse and some dirt. In the pocket of his green Ninja Turtles shirt I found three more rocks, the plastic cowboy who quite obviously belonged in the blue shirt with his horse, and, unbelievably, a small picture of a woman he'd torn from a USA Today.

I asked him about this last bit and he said, "Oh, yeah. Is she a supermodel? Sean says supermodels are babes." Is this what my son has learned about women? I wanted to weep.

Then I checked the pockets of Scott's shirts. I found a strip of rawhide, two handfuls of sunflower seeds, uneaten, an empty tin of mint chew, and a rock. No pictures of supermodels. Thank God. I showed him the rock. "Oh, yeah," he said, taking the rock and shoving into the pocket of his jeans, "this is a perfect skipping rock. I meant to show Joseph."

Is it just my family or are all men packrats? Anything shiny enough or smooth or smelly, anything that's shaped like a cloud-animal or looks like Mom might put it on the mantle, anything, basically, small enough to get in there, gets crammed into their pockets and brought home to the washing machine.

Shirts are new. I usually don't have to check shirts. Not in the winter, anyway, when shirt pockets are covered by smooth-fronted sweatshirts and dry-clean-only jackets. It's the jeans. Jeans are the mouth-pockets of these two humanoid, Huxhold packrats.

I started keeping track two weeks ago, storing and cataloging the stuff, like an anthropologist. When we're all long dead and some scientist digs up my laundry room, I want him to find something interesting. Not that our laundry isn't darn interesting on it's own, mind you.

Joseph's pile so far contains at least 40 rocks. Some are beautiful, with hunks of agate, or crystal chips. A few distinguish themselves by their lovely color, hazy purple or mossy green. But most of them are just rocks. Lava rocks and road gravel and big pieces of dirt he just mistook for rocks. He loves them equally.

Try chucking them back out onto the driveway someday. THEN you'll hear some frantic yelling, boy.

He also has a baseball card, three empty coin wrappers, a blank check from my checkbook (which would impress the supermodel, no doubt) a fistful of Legos, gum, more dirt, two politically incorrect bright red plastic Indians, nine tiny cars and a big white chicken feather from when the dogs got into the coop and ate a hen.

Scott's pile is more esoteric. He has a railroad spike, thirty or so drywall screws, a cap to the binoculars, a plastic connector to the electric fencing, a gear-shift cover, the dried-up sack from a castrated bull calf (no kidding), a broken metal ball hitch, half a green glass insulator from a power pole, six bucks in change and my kindergarten picture, which he shows to people and says, this is my wife.

Neither packrat carries a wallet. No room, apparently.

I showed them the stuff today at breakfast. They were thrilled. "Ooh, so that's where this was," and "Mom, I've been LOOKING for that!" They scrabbled through every little piece. The trash bag at their feet stayed flat. Their pockets, on the other hand, bulged. I expect I'll be seeing everything again when wash day comes around.

CHAPTER 2
COWSPEAK

Cowbonics: Let's celebrate Idaho's other language

By Steve Crump

Been following the debate in Oakland, Calif., over whether the public schools should teach street English?

Ebonics, it's called, and the Oakland School Board last month declared it a second language, worthy of pop quizzes and notes to parents.

("Dear Mrs. Rogers: Johnny was sent home from school early today after he was caught reading Shakespeare in Ebonics.")

Hey, what's good for Oakland is good for Oakley. I'm going to write Anne Fox and tell her that it's high time our kids learned Cowbonics.

Cowbonics is, after all, the language of the streets – OK, the muddy backroads – of south-central Idaho, as important culturally as a free lunch on a weed tour.

Test your Cowbonics IQ:

(Choose one)

1. Which is the correct use of the verb "to see?"

a. I seen

b. You seen

c. We seen

d. Obseen

e. All of the above

2. How may syllables does the expletive "shoot" have?
a. One
b. Two
c. Three
d. One "she" and unlimited "oots"
e. None of the above

3. Which of the following do you get your truck stuck in after one too many beers?
a. Borrow pit
b. Barrow pit
c. Ditch
d. Second gear
e. Some of the above

4. Which of the following terms is not considered cussing in eastern Idaho?
a. Flip
b. Oh My Heck
c. Sucky
d. Dagnabbit
e. Not allowed to say in mixed company

5. Identify the correct pronunciation of "barbed wire":
a. Bobwarr
b. Brbwurr
c. Burbwor
d. Climb over the fence
e. With an Angus bull in hot pursuit

6. Which of the following would you expect not to find in Fart Hall:
a. *An argandy farmal*
b. *A harse* trailer
c. *Arion* Samuelson
d. *Marse* code
e. *Aral* Roberts

7. Which of the following doesn't rhyme with the word "hill?"

a. Peel
b. Meal
c. Kill
d. Steal
e. We'll

8. Who controls 70 percent of Idaho?

a. *Fedrul gummint*
b. *Fideral gummint*
c. The federal deal back there in *Worshington*
d. Bruce Willis
e. The *nukulur inergy bidness* on the federal deal out *Orco* way

9. Which of the following is not a synonym for "rude?"

a. *Ignurnt*
b. Real *ignurnt*
c. Peeved
d. Affected
e. Spud buyers

10. What is the capital of Idaho?

a. *BOY-see*
b. *BUOY-z*
c. Salt Lake City
d. Micron Technology
e. Greater San Fernando Valley U-Haul Center, Van Nuys, Calif.

Answers: 1. (e). 2. (d). 3. (e). 4. (e). 5. (c). 6. (a). 7. (c) It's a two-syllable word, pronounced *"KEY-l."* 8. (d). 9. (c) To be "peeved" is to be seriously annoyed either by an in-law or a member of your canasta club. 10. (e).

How'd you do? If you got six or more right, you're qualified to be governor. Four to six shows potential, but you need to hang out in the parking lot at a few livestock sales – thumbs in your belt loops and leaning on your pickup – to absorb some culture.

Three or less? That means you talk real purdy English. Better smile when you say that around here, stranger.

Young buckaroo strides toward manhood

By Suzanne Huxhold

CLOVER CREEK-Testosterone is a funny thing. And I don't mean funny peculiar. I mean funny ha-ha.

My son, at 7, already has the full complement of testosterone. As much as he'll ever need to run a ranch, anyway.

Never was that clearer to me than at our branding this weekend.

When our friend Tim, a Colorado cowboy, arrived Thursday, Joseph went immediately into full hero-worship mode. Wherever Tim went, Joseph followed. When Tim put his finger aside his nose out in the hay yard and discharged himself of some alfalfa dust in a truly masculine manner that women would rather die of sinus blockage than employ, Joseph decided forever after that Kleenex was a girl thing.

When Tim spit, Joseph spit. When Tim tied baling twine into a loop and tried to rope the dogs, so did Joseph. I watched in wonder when, after Tim gripped the edge of his cowboy hat with thumb and forefinger and bid me a courtly good morning, my son imitated him perfectly.

Joseph came charging into the house at about 8:30 a.m. The cows had been fed and the men were planning to move a bunch of first-calf heifers into the corrals.

"MOM," he yelled, even though I was two feet away, "I need my chaps."

"Your chaps? You don't need your chaps, honey."

He looked at me as though I was the stupidest female on the face of the earth.

"I NEED them, Mom. I can't RIDE without them."

I saw this was important to him, though God knew why, so I didn't mention he'd been riding sans chaps for six years now. I buckled him in and he dashed out. I looked out the window and noticed Tim was buckling on his own chaps. Ah.

A few minutes later, I dried my hands on my apron and walked out to the barnyard to watch the men saddle up. Joseph sat atop his horse in the center of the yard, saddled but not bridled. I grabbed his lead rope and started him out the gate.

"Mom," he hissed at me. "Put my bridle on."

"That's O.K., honey. I'll just lead you out."

"Mom," he sounded pained now, spacing his words carefully through his clenched teeth and looking fretfully over his shoulder at the ten cowboys saddling up behind him. "Put - my - bridle - on- NOW!"

I led him out the gate and put the bridle on his horse for him. He grabbed the reins from my hands. "Open the gate," he ordered.

When I did, he kicked his mount back through, rode around the cowboys once or twice to prove he wasn't the type of weenie who needed his mother to lead him around a barnyard, said "Hey, Tim," in his deepest, most nonchalant voice, and sauntered back past me without a glance in my direction.

He rode all day, his father told me later, without a whimper, yipping at the cows, scolding the dogs, calling his horse a "dirty old rip" when he wouldn't ford a ditch properly. Scott told me his son looked up at him once during the long day, cocked a skinny arm against the sun, and said, "You know, Dad, Cupcake is a baby name for this horse." His father agreed with him very solemnly, but smiled when he recounted the story to me.

Dinner cooked, dishes done, I went out to the hitching post that evening. Scott, Chuck, Bill and Tim were discussing the day, discussing the cows, discussing the grass. They each had a fat bottom lip as reward for their 18-hour day, and punctuated each sentence by expectorating a long brown sluice of tobacco juice into the gravel.

I looked at my little son, who it seems just weeks ago was diapered and sweet-smelling and sitting on my lap playing itsty-bitsy spider. He was crouched in the dirt, picking at the manure on his boot with a sharp stick, stopping at intervals to pull on his Mountain Dew and then, to my utter astonishment, spit it straight back out through the gap in his mouth where his two front teeth used to be.

He must have felt my regard, for he looked up then and squinted at me. I couldn't help myself. I leaned over and kissed the top of his sweaty head. He glanced quickly up at Tim, making sure he hadn't witnessed the revolting display.

"Mom," he said, disgusted. "I'm not a baby."

Oh, Joseph, I know. I know.

Crabby cows and contentious cowboys

By Suzanne Huxhold

CLOVER CREEK-You know, I'm surprised there aren't more double homicides in the cattle business.

No, really. The fact that these guys don't murder each other with more regularity is a shock. If there is a more hot-headed, hard-headed, thick-headed group of people on the planet, I'm glad I don't know 'em. Bosnians, maybe. Maybe the Bosnians have it over cattlemen, but it'd take some proving for me to believe it. I'd actually have to be shelled, I imagine, to be convinced there are crabbier men than the guys I work with.

We were sorting cows in the gulch this week. Well, actually, the men were sorting cows and I was hauling food from home to the hills. I got to the sorting meadow about 11 a.m., the back of the Jeep loaded to the struts with hamburger and chips and plastic chairs. I didn't have my glasses on, so I jogged across the dirt road to where my husband was sitting, atop his horse. The glasses I mention because if I'd seen the scowl on his face, I would have stayed where I was, doing my women's work quite happily, thank you.

"How's it going?" I asked.

He just grunted.

I scanned the herd. Actually, the three herds and a dozen or so strays they had pulled off the hills over the previous two days.

"Looks like you're making headway."

He glared at me. "Is that what it looks like?"

"Um..."

"'Cause if that's what it looks like, then you need your glasses." Of course, I had already figured that out.

He then proceeded to complain, as cattlemen will, about the stubbornness of the cows, the recalcitrance of the weather, the inflexibility of the cowboys.

"Oky-doky," I said, backing away, "I've got to get dinner on."

"And THEN," he continued, "your damn HORSE..." I left him at a run.

By the end of the afternoon, when all the burgers had been eaten and most of the beer consumed, I'd heard it all. Every complaint.

"...why, they stood with their backs to the herd all morning..."

"...in a bunch, just talking like they had all the time in the world, while we was running our horses dead just trying to keep them cows separated..."

"...one shell'd killed 'em all..."

"...damn stubborn. He's half rattlesnake, if I didn't know any better..."

"...had them cows so turned around, you'd a thought they was heading for Canada..."

"...gonna kill that poor cayuse, running him like that..."

"...Dammit! I TOLD them kids to keep that bunch separate..."

"...you'd think some of us might have nothing better to do than clean up the messes of another..."

I just kept shoveling food in 'em, hoping the fat meat and cold drinks would take the edge off their tempers. It didn't surprise me, all the complaining. For one, working with cows for longer than about 10 minutes is bound to turn a person surly. And second, I've yet to meet a single cowboy who didn't think he was the boss of the world. Or should be. It's natural. Why, a sane man wouldn't choose this life. A sane man would get a 9-to-5, with employee benefits and a regular paycheck and someone else to worry whether all the ducks are in a row. Or all the cattle are at the ranch, whatever. By virtue of his career choice, you've got to assume every cowboy is a iconoclast. And it follows that, as one iconoclast is bound to be trouble, 15 of 'em trying to separate one brand from another on a hot dusty hill in late June is something like a bunch of, say, BOSNIANS, trying to run the U.N..

On the other hand, iconoclasts tend be charming as hell. Which is why, once the work is done and they're off their high horses, cowboys band together like drunken Irishmen, trading stories, laughing like lunatics, praising one another's horsemanship.

"...you know, I'd a lost that one old biddy if you hadn't stopped her..."

"...hardly believe you was able to catch that calf. Fine work. FINE work..."

"...always a pleasure to have good hands on a day like this..."

I'll take that Ghengis Khan breed

By Suzanne Huxhold

CLOVER CREEK-Scott went up to Blackfoot last week for a bull sale. Bought a beautiful bunch of Simmentals. The breeders delivered them yesterday, and now they sit out in the muddy corral, placid behemoths, mouthing cud and blinking at each other as though they haven't a care in the world. Which, being bulls, of course they don't.

The night of the bull sale, my folks and I went to dinner with cowboy Joe and his wife. When I told Joe that Scott was in Blackfoot buying Simmental bulls, he said, "Good. He ought to just stay there."

Of course, I knew what he meant by that. He meant since Scott was buying big-headed, breachy bulls that throw calves that get hung up at the ankles and turn into poor mothers who fight the horses and give skim milk, my husband ought to just stay in Blackfoot and check himself into the psychiatric center there. That's what he meant by that.

Joe is like every cowboy I know. Every cowboy I know sets his sights on a particular breed sometime early in his career and can't be dissuaded from the idea that his breed is superior in every way to all other breeds; the very pinnacle of cowness, the litmus by which all other cows are tested.

Joe's favorite is the "red-necked" cow. They are called red-necked not, as you might be thinking, because they watch pro-wrestling in their underwear, but because they have the white faces of a Hereford and the red bodies of a Red Angus. Joe likes this particular cross above all the myriad other cattle crosses because red-necks have reasonable-sized calves, are excellent mothers and have, for cows, what can be considered good dispositions.

Though, frankly, that's a bit like saying Ghengis Khan had a good diposition, for a Mongol killer.

But Joe's opinion on the subject is only his own. One would think all cowboys would want a reasonable-sized calf with a good disposition, wouldn't one? But one would be wrong. There's a wagonload of breeds out there, and if you have an open mind you come to see that every breed has something to recommend it. If you have an open mind, you can also end up with a calico herd, which is one big reason you'll never meet a cowboy with an open mind.

Black Angus cowboys say their breed produces a higher-grade carcass. Limousins, according to their supporters, have sky-high wean rates and a very cool name. Saler calves slip from the birth canal like butter, Charolais steers dress out well and Brahmas could find enough to live on in a rock pile. Watusi cows have, well, really big horns. Herefords will mother anything in a calf suit. Longhorns, as any Texan will tell you -and tell you- live forever.

This isn't just soft-headed favoritism, mind you. Breed boosters are fanatics. Like a crazed bunch of beauty contest judges, they sit at livestock auctions and county fairs and compare the near mythical strengths of their favorite against the miserable weaknesses of the opposition.

Too much fat, they point out derisively, or too lean. Thick ankles. Wide hips. Never have liked the look of a Braunvieh, gimme one of them old Shorthorns any day. Shorthorns? Why I wouldn't take a shorthorn for a walk. Now take that Simmi over there. THERE'S a cow for you.

If the cows had an evening gown competition, they could put it on TV. Well, cable anyway.

I once sent an article to a publication in the Midwest; an article on the vagaries of the fall calf market. I got a rejection notice from this publication because, they told me, they only accept articles that deal specifically with the Brangus breed. The BRANGUS breed. In order to sell anything to these people, I'd have to write specifically about the lop-eared, hump-backed Brangus breed. Nice cows, I'm sure, probably the best around. I would be hard-pressed to write an interesting paragraph about 'em, though, much less a whole article. But these guys publish a monthly magazine, just crammed with great stuff about only the Brangus breed. This is just what I mean when I say cowboys are fanatic.

Scott, thankfully, is not a fanatic. He prefers to think of himself as a scientist. All the data isn't in yet, so every year he tries something new, just to see what he gets.

Our bull pasture is a Rainbow Coalition's dream. We have Longhorns out there, and Herefords, Black Simmentals and yellow Simmentals, a Saler, a Red Angus and two big old cross bulls I can't quite identify but that Scott held over from his herd because of how well they kept with their cows.

Scott's betting the combination of bloodlines will someday make him a millionaire. On second thought, maybe he should have stayed in Blackfoot.

CHAPTER 3
NEANDERTHALS, OR NIGHTS AT THE TUTTLE MEN'S CLUB

Neither rain nor cold nor dark of night

By Suzanne Huxhold

CLOVER CREEK-My husband attended a Bassmasters Tournament at Brownlee Reservoir this last weekend, the first of the season.

The buildup was phenomenal. Really. Just amazing. A whole winter's worth of fishing frustration and anticipation was riding on this first tournament. (More than that, though, now that I think on it. A whole lifetime's worth. Ask Scott what his life's dream is and he'll tell you he wants to get himself on the Pro-Bass Tour. Go ahead, laugh. But he makes his living ranching right now, and what could be more pie-in-the-sky than that? If ranchers were entirely realistic people, they'd be accountants.)

Anyway, Scott and my Uncle Bill had some serious enthusiasm going at our house Thursday evening. They were giggling like schoolboys and comparing leeches and promising Aunt Marcie and I they'd return to us conquering heroes, fish on line, trophy in hand.

They left Friday afternoon, intending to spend the night in the van so they could get an early start Saturday morning. The weather Friday was clear and calm. Spirits were soaring. If a person could get on the Pro-Bass Tour on zeal alone, Scott's dream would have come true that day.

They didn't pack any food, did I mention that? They meant to get groceries in Cambridge. They overestimated Cambridge.

They ended up eating peanut packets and cheese and breadsticks they found at the checkout counter at the Grange.

And they forgot to let one another in on the snoring thing.

Buzz saws? Buzz saws WISH they could make as much noise as these two when they sleep. If one dozed momentarily, the other awakened immediately. All night long.

Plus, the weatherman lied. Lied like a rug. The little squall he was looking for on Monday hit Brownlee like a monsoon on Saturday. It poured. The heavens opened. Noah didn't see that much rain. And Scott and Bill, in the whole leech-comparing frenzy of

Thursday, forgot to pack raingear. Scott had his insulated coveralls, but they weighed 70 pounds by 9 a.m. and he couldn't lift his arms to cast, so he ditched them.

Starving, sleepless, soaked and freezing, they fished. As though the world might have come to an end if they hadn't.

When mild hypothermia set in, and they started getting the shakes, they had trouble changing worms, and spilled an entire bottle of fish gunk on their clothes. When they caught fish, it took twice as long for their frozen fingers to pry them off the hook, and because they couldn't see very well in the downpour, two of the five they thought might win them the trophy were woefully undersized. Guppies, basically. The boat filled with rainwater up to their ankles and shorted out the live well. A log dislodged from the bottom floated to the surface with the force of a bullet and broke off a piece of their new prop.

When they got home, after a 4-hour drive, they were still wet.

I heard the details while wringing out Scott's socks and underwear.

"Why didn't you just come home?," I asked, in that female, pragmatist sort of way I have. That come-in-out-of-the-rain, don't-stand-in-water-while-operating-electrical-equipment, when-your-fingers-freeze-together-it's-time-to-get-back-in-the-car mentality that's so foreign to my husband and sportsmen everywhere.

"It was a TOURNAMENT," he told me, in a way that made it pretty clear he thought I was too stupid for words. "We weren't just FISHING!"

The funny thing was, he said, while sitting in a tub of hot water, trying to rub some feeling back into his rear end, was that their two best fishing buddies stayed in a motel Friday night and didn't even get to Brownlee until Saturday morning and THEN left for home before it even got dark Saturday night. He was just mystified by that.

"Yeah," I said, trying to lift his fish gunk-covered coveralls into the washing machine with a winch. "Foolish them."

Passin' time at Whisky Mike's

By Steve Crump

"I asked a very great man the secret of life. The secret of life, he replied, is knowing where the fish are biting.

"So I asked him the secret of knowing where the fish are biting. He was not precise."

— Mark Twain

When the first wagonloads of pioneers rumbled through the Portneuf River valley along the Oregon Trail, it was said, Whisky Mike was there to sell them fast talk and slow mules.

Nobody really knew how he got there or whence he came, but in the afterthought of a valley that old Highway 30 split like a rotten log, his clapboard store and corrugated metal garage patched with flattened maple syrup cans were all there was to the settlement called Pebble.

Pebble was named for a step-child's portion of a creek that dribbled out of the mountains, but it was a real town all right — as real as the tumbleweeds that embraced the Burma Shave sign.

It was one of those whistle-stops along the UP where there was a lot more whistling than stopping, and its residents consisted of Mike and his ancient cocker spaniel, Taffy.

The valley was just an interlude between Lava Hot Springs and Bancroft, but only the long-haul truckers made up time here. Everybody else stopped at Mike's.

Why is problematic. Mike was a vaguely disagreeable fellow who shaved last week, favored dirty jokes and T-shirts to match and wore his belly over his belt buckle. His combination saloon and store wasn't clean, and anybody who wanted a can of corn off the shelf to chum trout in the Portneuf had to blow the dust off first.

But it was one of those places with a mysterious allure, or at least mysterious lures.

Across the highway and over the barbed wire fence, the Portneuf was in no hurry to get to Pocatello. It slipped in quietly from the east, out of Chesterfield Reservoir, and ambled through pastures and alfalfa fields until it ran out of room and ducked through a gap in the mountains like a cat burglar with promises to keep.

At its deepest, the channel was six feet wide, and except when it flooded, the water was never more than two feet deep. But in its shadowy eddies and black pools, long, lazy rainbow trout wasted the summer.

You could catch them, it was said, with an angleworm from your garden or a red-and-white spinner from Skaggs Drug — but not often. The trout, you see, were Mike's.

Mike trafficked in hackle and tackle — salmon eggs the color of fire-sale polyester and nightcrawlers the size of your little finger — but most people sought him out for the unconventional.

Cheez-Whiz, for instance.

Mike had a tried-and-true method of making processed cheese product appeal to trout: He left it on a shelf in the stock room for five years. By the time you pried the lid off the jar, it would make Taffy's eyes water, set him to howling and send him fleeing into his hiding place in the cedar breaks behind the store.

But it was absolutely irresistible to the trout of Pebble.

One time my neighbor, Ralph, was fishing off the highway bridge and leaned on the railing on which he'd left a jar of Mike's trademark spreadable cheddar-with-chives. When he reached up to swat a mosquito on his neck, he knocked the jar into the river with his elbow.

Within seconds, the water beneath the bridge began to boil with trout, and by the time the jar floated ashore, even the label was gone.

And on the day Mike sold his last jar of Cheez-Whiz, the trout stopped biting.

The highway department eventually rerouted U.S. 30 over Fish Creek Summit to the south, and the only anglers who turn up at Pebble nowadays are fishing for nostalgia.

On the side of Mike's garage, the Carling Black Label sign that once showed a fisherman hoisting a trout has faded until only the angler remains, his arm raised as if beseeching the gods who used to live in these parts for one last perfect summer's day.

But if you're lucky enough to find that day and linger until twilight yields to dusk, you can sometimes catch the glint of a silver fin beneath the Pebble Bridge and a whiff of memory.

And from the cedar breaks, Taffy is howling.

Admire your neighbor's dead animal month

By Suzanne Huxhold

CLOVER CREEK-October is Admire Your Neighbor's Dead Animal Month in Idaho. We've been besieged by celebrants for the last 30 days.

Because ours is the first ranch you come to when leaving Unit 45 with a recently-plugged deer hanging out the tailgate of your pickup, everybody we know naturally feels obligated to come by and show off their kill.

To tell the truth, though I'd rather eat venison than just about anything (don't tell the Beef Council I said that), I haven't been altogether thrilled with this fall ritual before now. Maybe I was too soft-hearted. Or maybe it was the flies, I don't know.

I've had a change of heart, though. The ritual has patterns I'm starting to appreciate in my middle-age, a paradigm steeped in ceremony.

Admire Your Neighbor's Dead Animal Month starts the same way every year, with a casual visitor driving slowly in circles around our house one evening in early October, peering into the windows and trying to look blasé, until Scott comes out the back door.

"Been huntin'?" my husband will inquire needlessly, crossing the driveway to lean into the passenger window of a dusty pickup.

Of course he's been huntin'. We only see these people when they've been huntin'. It's a given. We wouldn't recognize them at our own dinner table unless they were wearing desert fatigues and a set of rattling antlers tied to their cartridge belts. But the asking is part of the ritual.

"Yeah," the hunter answers nonchalantly. Another part of the ritual of Admire Your Neighbor's Dead Animal Month is that the hunter is obligated at the outset to act as though this mountain of meat and bone simply fell into the back of his pickup with no effort on his part whatsoever.

"Get anything?," Scott asks casually, ignoring, for the moment, the gigantic, antlered cadaver lying in the bed of the truck, eyes staring open, rigid legs hooked over the sidewalls.

Now, I can see the thing from my kitchen window, but Scott pretends not to notice it until it is presented to him. It's polite, I guess, like waiting for your hostess to pick up her fork before you start eating.

The hunter doesn't deign to answer such a foolish question as this. He simply jerks his thumb toward the back of the truck. Scott leans back and studies the kill.

This is where things get tricky in this particular ritual. If Scott expresses too extravagant an interest he runs the risk of insulting the hunter by insinuating he's surprised the big dolt has actually managed to shoot anything. If he appears indifferent, the big dolt will know Scott doesn't respect the proper terms of the unspoken, ancient agreement between men under which all this kind of stuff - huntin', fishin', beer drinkin', shootin' at cans and peeing communally - falls, and when Scott drives his dead deer around and around this guy's house, he won't come out the back door at all.

But Scott is perfect. He grunts and stares appreciatively, asking pertinent questions. The hunter warms to the subject and waxes on about tracking and antler points and gutting and skinning.

When the bored look in his eye is fully transformed into a fevered, caveman expression of glee and adrenaline, Scott motions for me to bring out the beer. Then they stand in the driveway, drinking beer and discussing every animal they've ever killed or just seen dead on the side of the road someplace for up to, but not exceeding, an hour. (After all, the hunter has many more neighbors to pretend to encounter casually, and many more beers to drink, before he can haul that sucker over the rafters of his barn by its back legs with a length of baling twine and call it a season.)

It's kind of exciting to watch, this ritual of admiring the neighbors' dead animals. It makes me feel a part of some ancient cycle, a rite, a bringing to the village fire fresh meat and stories of the hunt. In accepting the ritual I feel I've brought myself much closer to an understanding of the male members of my species.

Without actually having to pee with any of them.

How do you tell if you are a real Idahoan?

By Steve Crump

The other day I got a call from a fella named Ted, who has lived in the Magic Valley for five years since escaping Bakersfield. Ted was anxious to know at what stage he could expect to achieve true Idaism.

That's a tough question, Ted.

Idaism is really less a state of grace than an attitude —- of having the confidence to know just how early in the winter to put that bale of hay into the back of your pickup, what social occasions call for screw-top wine and just how big a belt buckle you need to be considered overdressed.

But I've put together a little checklist for you that I hope will be helpful.

If you agree with more than 15 of the following statements, call me. If you score less than 10, call Jerry Brown.

1. I do my best napping at traffic lights.

2. A set of new mud flaps makes my palms sweat.

3. I consider the invention of the 8-track tape and the dashboard beer-can holder to be the pinnacle of human progress.

4. Brown gravy is the staff of life.

5. Given the choice between having a root-canal or getting an all-expense-paid trip to Chicago to see the Oprah Winfrey Show, I'd go to the dentist.

6. I don't know anybody who voted for George McGovern.

7. I have sometimes worn bowling shoes on formal occasions.

8. I would not drive my wife's car to a dogfight.

9. I consider ketchup one of the four basic food groups, along with beer, chicken-fried steak and nachos.

10. I have more than one pickup body in the back yard.

11. Garth Brooks seems like a nice feller, but I don't trust guys who wear doubled-breasted shirts.

12. "Thunder Road" was a better movie than "The Sons of Katie Elder," but only because of Mitchum.

13. When I get depressed, I go to D&B Supply and buy Rustoleum in bulk.

14. I consider capital punishment by lethal injection a slap on the wrist.

15. My bloodhound, Festus, can beat up your golden retriever any day of the week.

16. When I heard the Idaho Legislature had designated a state fossil, I thought they were talking about themselves again.

17. Light beer is the surest sign of the decline of Western civilization.

18. I wore a seed-company cap instead of a mortar board at my high school graduation.

19. Boise is what happens when you let yogurt stands and yuppie bars into cowboy country.

20. I can spit snoose from a moving pickup and not end up wearing it.

21. When everybody smoked Camel straights, the Japanese owed *us* money.

22. It's a lot more important to wash and wax your horse trailer than to wash and wax your horse.

23. The federal government is the last, best evidence that Benjamin Franklin slipped a whoopy cushion onto George Washington's chair before he sat down to sign the Constitution.

24. I don't take my cowboy hat off at the dinner table and I don't take my cowboy boots off in bed.

25. The California border runs through Ketchum.

26. I buy my shaving soap by the cake, my Aqua-Velva by the quart and my Brillcreme by the barrel.

27. I have attended the weddings of many of my cousins —- to each other.

28. I don't see anything funny about Ross Perot's haircut.

29. I believe America needs to get back to basics, and redress every citizen's inalienable right to life, liberty, set-asides and FmHA loans.

30, You'll have to pry the remote-control to my big-screen TV from my cold, dead fingers.

Opening day

By Suzanne Huxhold

CLOVER CREEK — "It's opening day of fishing season Saturday," I remarked to my husband, the bass fisherman, yesterday. "Huh," he answered noncommittally.

"Well," I prodded, "isn't that exciting? I mean, opening day.

That's like some kind of high holy day for you guys, isn't it?"

He rolled his eyes a little and gave me a pitying look. "I'm a Bassmaster." As if I didn't know. As if I don't wash his Bassmasters shirt and hang up his Bassmasters hat and talk to all his Bassmasters friends at the annual Bassmasters Christmas Cotillion at the Royal Lounge.

"So?" I said. "It's still opening day."

"Honey," he explained as slowly and patiently as he could, which, being as he's a man and all, is not too slow or too patient, let me tell you, "I fish for bass. Bass live in lakes and reservoirs. The lake and reservoir season is open year 'round."

"So what's the point of an opening day if you can fish all year?"

"Opening day is for stream fishermen."

"And stream fishermen don't fish for bass?"

He only looked at me, wondering, I'm sure, how he ever got stuck with a wife as dumb as me.

The simple truth is, I've never understood it. Any of it. Any of the get up at four, drive three hours, sit all day, staying-damp-shooting-bull-staring-at-the-water-till-you-go-blind appeal of fishing. You mean you don't take a book along? I remember asking my husband when we were dating. A portable TV? A radio, even? No, he told me. Never.

But the way I don't comprehend the timetable of fishing or the mystique of fishing is nothing compared to the way I don't understand the politics of fishing. I was chatting with a friend of mine the other day. He's an avid fly fisherman. The kind who makes his kids sleep four to a bed so he can have a whole room in his house just for feathers and colored string.

"Hey," I told him. "The new principal up at the school is a fly fisherman. You guys should go fishing together sometime."

He grimaced. "He's a float-tuber."

"Ah," I said knowingly. "the scum. What does that mean, anyway?"

"Well, float-tubers," and he said this with an almost imperceptible sneer, "throw wooly worms. Dry- fly fisherman try to match the hatch."

Uh huh. Well, THAT certainly explained it. I got the gist, though.

Apparently, fishing is like war. Once you choose sides, there's no going back. And everybody thinks their army is the smartest, the toughest, the best.

Dry fly fisherman, evidently, think float-tubers are wooly worm-throwing weenies. Float-tubers, from what I can gather, think spin fishermen are slack-jawed morons. Spin fishermen think worm fishermen are the very minions of Satan. Worm fishermen think everyone else has debauched the purity of the sport with high-tech gadgetry, which according to lure fishermen is akin to saying childbirth has been absolutley RUINED by that whole hand-washing idea. And Bassmasters, and this I know from personal experience, think all fishermen besides them are wasting their time even leaving the house in the first place, since bass are the only fish that really fight, and anyway, Bassmasters have huge engines on their boats, HUGE engines, and what's the point of fishing if you don't have a really loud boat?

I've decided to be Switzerland about the whole thing. Being as I'm apparently the one person left in Idaho who thinks the only time a civilized person should actually see a fish is after it's been sauteed in butter with maybe some nice shallots, and not swimming around in a live well with a hole in its mouth where a hook used to be, I think that's my privilege.

Chapter 4
The Tropic of Hagerman

Idaho winters and other tall tales

By Steve Crump

I was 8 years old, helping my mother dig our '56 Chevy out of a snowdrift in Arbon Valley, a farming area southwest of Pocatello.

It was late February, and it had snowed off and on for three months. The snowplow that normally cleared out our country lane had been stuck in the borrow pit for about six weeks, by now buried to its cab.

The road to town had been closed for almost as long, so the only route out was over the Garden Creek Gap — the long way. We'd tried it on this particular day, and gotten nowhere.

The car had skidded and fish-tailed, ending up with its nose through a barbed-wire fence and its back tires squarely in the middle of Rattlesnake Creek, where they broke through the ice and settled into the frozen mud below.

"Well," Mom said, as she shoveled away. "At least it's not as bad as the winter of `49."

Ah, the winter of '49. When I was growing up, its legend loomed larger than the Great Depression, the Spanish Inquisition and Noah's Flood combined.

All activity, I was led to believe, stopped colder than a farm-loan officer's heart.

Cattle by the millions made good their escape over fences buried in snow, Utah Power linemen chipped ice off frozen lines without climbing the poles, and elk and deer had to take refuge in Paul Bunyan's footprints.

Paul himself hadn't been seen since Christmas, but his footprints led south.

The lesson was clear: You live in Idaho, where winter isn't to be trifled with.

Yeah, well.

Having spent 36 of my 44 winters here in the High Toasty, I'm about convinced that we've all been jackaloped on this subject.

You remember the jackalope, don't you?

It was the star of a series of picture postcards back in the '50s and '60s. Using dull scissors and too much paste, some Idaho photographer grafted the head of a jackrabbit onto the body of an antelope and sold the cards to gullible tourists from California who had too much money.

The astonished visitors would buy the cards by the sackful, mail them back home, and then head out into the toolies in search of these marvelous creatures.

There they'd get lost in some farmer's 300 acres of stinkweed, run out of gas and have to hike back to town.

The boys at the gas station would gather round Cousin Ed's Ford pickup and laugh so hard that Uncle Dube would get the hiccups.

Well, somewhere, I suspect, Uncle Dube's having a belly laugh over the winter of '49 at our expense.

There's no question that them folks had snow back then — I looked it up. But according to the Weather Bureau records, the precipitation in Pocatello, Idaho Falls and Twin Falls during December, January and February that year was less than the average annual snowfall in Dubuque, Iowa; Decatur, Ill., or Billings, Mont., and the temperature was warmer.

I think we been bamboozled by a bunch of cold-weather wimps.

I thought so three years ago when all those storms paid a rare visit to our little corner of the desert. Some people even had to buy snow shovels.

Could it be that Idaho is to winter what Velveeta is to cheese?

You bet your overshoes — the ones you haven't had to wear since fourth grade.

They don't call it Sun Valley for nothin'. The dirty little secret of our rugged, wind-buffeted steppe is that we grow wine grapes here.

Winter of '49, indeed. Shoot. Probably had to wait to get a tee time.

It's high time we all admitted that the only difference between Idaho and Arizona is that the cheat grass grows thicker there.

I'm gonna take my insulated coveralls, my 200 pounds of rock salt and my snowblower back to D&B and see if I can swap them for a new air-conditioner pump. Probably need it by next week.

Unless, of course, it snows.

Rain

By Suzanne Huxhold

CLOVER CREEK — I went up to the Ox Bow Cafe this morning. I had to be in town early anyway, so I stopped in for breakfast and coffee.

There's nothing like a cowtown diner to gauge the mood of the community. Dry days and stack fires and drops in calf prices can be read on the face of every man hunkered at the counter. Bad times in the business make for quiet mornings at the Ox Bow, and the farmers and sheepmen and buckaroos keep their hats pulled low over their eyes and their misery company.

But you should have seen them today. You should just have seen them. Whatever lingering gloom that had settled over this town during nine straight years of unrelenting drought was rained away yesterday. And they were celebrating up at the Ox Bow.

I walked in at 8:30, shoes squishing a little and water pouring off my coat. And stopped. And stared. Every seat at the counter was full, the boots at the foot rail damp, the rain-spattered hats pushed back from weathered faces. And those faces, yes they were! were smiling. Idaho desert farmer smiles; content, for now, a little restless maybe, and amazed.

A friend of my husband's looked up from his steak and eggs, pushed at his hat until it rested square on the crown of his balding head, and laughed at me. I tried to remember if I'd ever seen his teeth before. Or his forehead.

"Lovely weather we're having!" he joshed. I smiled, giving the correct response.

"For ducks and farmers," I replied, earning a hoot of approval.

Rain.

Everywhere I go these days, the dour looks and grim faces of the drought are temporarily transformed. Oh, they predict bad times will come again, farmers being farmers and all, but for now the prevailing mood is one of gratitude. They'll worry about wool prices and hollow heart and leaf scourge later. Next week, maybe.

When it's dry again. For now it's at long last, rain. Rain enough to soak to the roots of the apple trees and flush out the mice burrows. Rain enough to hit spring-grazed bunch grass and make it promise fall bounty. Rain enough to drench the alfalfa fields and make the sellers of diesel fuel and replacement sprinklers curse the leaden sky.

Rain.

I greet people I haven't seen since the last strong spring, two years ago. Fields too saturated to work, shop chores done and done again during the long winter, and nothing to do at home but surf through the satellite channels trying to find a rerun of Fishing the West, they come to town in a swarm of pickup trucks, as though washed down a river like so much feed store-capped, Ford-driving flotsam, to land shoulder-soaked and pocket-heavy at D&B Supply or Globe Seed; to rummage through the irrigation boot bins at the Farm Store and admire the chukar chicks at the co-op; to gaze out at the ponderous fall of water outside the window in the kind of heart-heavy wonderment only an Idaho desert farmer can understand.

Rain.

It's foolhardiness and ego that sustain us through the drought years. We dry up like our land, get all leathery and stiff-backed and sour-eyed. We pull more water from the aquifer and convince ourselves it will be replenished before it goes too far to ground.

We oil our pumps, set our wheel lines, drop yet another pipe into the river. And ignore the dust that begrimes our tongues and our clothes and our spirits whenever we step from the house.

Until some wet, miracle spring such as this reminds us our souls are as much at the mercy of mother nature as our land is, and that we can postpone her impact only so long.

I hope I never have to go through another spring like 1994's, and the eight before it, where every day begins with a search for wet clouds and every night ends with a plea to the rain gods for, please, something. I will, I reckon. I've been in the business long enough to know that. But for now I won't think of it.

For now, I'll go up to the Ox Bow Cafe and eat steak and eggs and enjoy the tipped-back hats for all they're worth.

It don't start to get interesting 'til it's 40 below

By Steve Crump

This is the weekend I always dread.

On the first Sunday of November, my Cousin Oren drags me down to the Bear River for what is euphemistically called duck hunting.

Now, duck hunting is no such thing. The object is to sit in a freezing bog in the hours before dawn, surrounded by reeds and quacking like an adenoidal platypus until the bourbon runs out.

The first one to leave wins. Never mind the ducks, who are by now sunning themselves in the big pool at the Cancun Hilton.

Still, duck hunting is much prized by us macho, cognitively-challenged outdoor types for the demands it places on the true sportsman, who is somehow unable to grasp that he could buy a duck at Albertson's for about one-tenth of what it costs him to winter in this miasmal swamp.

As my Uncle Fred explained it to me, the best duck hunters discipline themselves to stay perfectly still even as their feet turn to tree stumps and their ears snap off in the freshening north wind.

Ducks, he went on to say, are very perceptive, and if they espy a hunter bundled up like a mohair bowling ball down there by the banks of the Big Slurpy, they get to laughing so hard that they're danged tough to hit.

The Bear River, which wends its way in and out of Utah like wavering sinner, is ideally suited for duck hunting because it's six miles up in the mountains and close to Bear Lake, the only body of water within 100 miles that any self-respecting traveling duck would give a second look.

So it's important for the hunter to station himself strategically in the tall grass, where the wind can whip the cattails into your face and where your electric socks can melt the mire beneath your boots.

Mastering the duck call is vital. You can buy one at any sporting goods store, but the most effective method of imitating the way ducks talk to each other is to catch bronchitis. Fortunately, this is rarely a challenge after sitting for hours in driving sleet.

You also need to make sure that you have a good vantage point for when your shotgun knocks you on your well-padded backside and across the ice like a furry hockey puck.

Above all, be sure to arrive in your duck blind at least two hours before sunup. These shelters are called "blinds" because you invariably fall into the river trying to find them in the dark.

Fully awakened by your bracing swim, you can now settle down to a leisurely, man-to-man conversation with your hunting companion, carried on just above a whisper:

("I can't feel my fingers.")

("Shut up and pass the sterno.")

("How do I pass the sterno if I can't find my fingers?")

("Well, dip your fingers in the Thermos and thaw them out enough so you can you can pass the damn sterno.")

The awful stillness before the dawn is supposed to remind you how close you are to God. When you go hunting with Cousin Oren, that's more true than you think.

"Stay here," he whispers. "I see a spot across the river."

With that, Oren tip-toes to the edge of the water, silently slips into his 11-foot aluminum boat, shoves it out into the river, and yanks the outboard-motor starter cord, shattering the silence like a belch at a cotillion.

Once across the water, he finds himself a stand of pussy willows and awaits the one or two ducks who haven't already canceled their reservations on Bear Lake.

When they finally approach from the north, I note with terror that the barrel of Oren's Remington 12-gauge is pointed about 20 feet over my head.

Two years ago, I stood up at the wrong time and actually felt hot birdshot falling on the ice-cold nape of my neck. Now I confine myself to hunkering in the mud, whimpering softly.

But it's always worth the wait. If Oren hits a duck, it always falls in the water, and his cocker spaniel, Daffy, is smart enough to want no part of retrieving it.

So Oren paddles out into the middle of the river in search of the unfortunate mallard, and with dazzling regularly, falls in himself.

That's why I always keep the keys to the truck. By the time he's hauled himself out of the drink and retrieved *his* cold duck, I'm home by the fireplace enjoying *my* Cold Duck.

But it's a dry heat

By Suzanne Huxhold

CLOVER CREEK — Something pressing has come to my attention and I must ask you all a question.

Why the hell do we live here?

Seriously. Why do we live here? Why did our ancestors decide this was an OK place to reside? Did they come here in the fall? Did they decide, in October, this was a nice, temperate place to stay, never imagining what the summer could be? And by the time the summer finally came, did they have crops in the fields and babies on the way and some other force of nature or commerce that compelled them to stick by their decision?

'Cause I'm telling you, I've never been this hot, and I want some explanations.

And I want 'em NOW.

The thermometer on my west-facing porch has pegged out at 110 degrees every afternoon for almost two weeks. The log walls of the little cabin by the pond absorb so much heat, the thermometer on its west-facing porch has hit 120 degrees. 120 degrees. Human beings aren't supposed to live at 120 degrees, unless they've been bad and couldn't get into heaven. I haven't been bad. I've only been somewhat crabby and slightly neurotic. But not bad. SO WHY AM I BEING PUNISHED?

I have a theory, if you'd like to hear it. OK, here it is.

I don't think our ancestors came here in October and set up housekeeping thinking the weather was always as nice as it is in the fall. I don't think they looked at all the black rock and blistered desert and thin streambeds and thought this was a swell place to stick a town. I don't think our ancestors were stupid.

I think our ancestors were evil. I think each of them, including Mr. Perrine and Mr. Addison and Mr. Blue Lakes and all the rest, were servants of Lucifer. I think they found the hottest damn place in the United States and decided to move here because they were COMFORTABLE in the bowels of hell. They LIKED it. They were, in all likelihood, the very spawn of the devil.

We're stupid. Too stupid to move someplace else. Like Alaska. Or Greenland.

There can be no other explanation, and if you sat long enough on my west-facing porch in the afternoon, the sun baking into your brain and scorching whatever tiny bit of reason you had left, you would probably come to the same conclusion. If I had an air-conditioner or a rental cottage on the Oregon coast, I probably wouldn't accuse the original settlers of this valley of being minions of Satan, but there it is. I don't have an air-conditioner or a cottage on the Oregon coast. I'm hot, and I'm mad.

In fact, I can't ever remember being this hot and this mad.

Last summer was positively bracing. The most pleasant summer I can remember. I don't remember spending a moment of last summer contemplating the sinister motivations behind the sojourners of the Oregon Trail.

And although the previous 29 summers of my life were not always spent in clear comfort - how well I remember the stench of a summer camp latrine on a warm afternoon - I can't remember anything like the last month or so of unrelenting heat.

The members of my Pacific Coast-based family haven't been very understanding. The settlers of their valley were winged cherubs compared to the settlers of our valley, at least in terms of weather. (We can discuss their Beelzebubs of traffic another time.)

"It's so hot," I'll wail over the phone.

"It may be hot, but its dry heat," they'll answer firmly.

"That must be why my skin is cracking off in big bloody chunks then, hmmmm?" I'll reply sarcastically. I have no patience for my family. The last two weeks have led me to wonder if I have patience for anyone anymore.

One golden moment has emerged in the midst of all this shining, simmering heat, I must admit. My mother was here last weekend. She stayed in the cabin of the 120 porch. Before she left for home and 85-degree weather, she dropped a check on my kitchen counter.

"Buy a swamp cooler," she said, her limp hair plastered to her sweaty head, rivulets of perspiration soaking her shirt.

"But Mom, "I said sweetly, "It may be hot, but its dry heat."

She might have glared at me. I can't say. I was already on my way to town.

Chapter 5
Safe Sex in Jerome County
IS A PADDED HEADBOARD

Men in shorts

By Suzanne Huxhold

CLOVER CREEK — I was canoeing with friends the other day, down a narrow stretch of the Henrys Fork of the Snake. Anne and I were in the lead canoe, Marla and another Anne ("The Ann's," we call them, to their undoubted irritation) were in the canoe behind.

We had planned the trip for months. We wanted to fully experience the local wildlife. We prayed for sandhill crane and osprey, perhaps a glimpse of elk or deer, even moose. We had our binoculars ready, our cameras loaded with high-speed film. It was a nature trip. An expedition of purest intent.

But when we came upon an isolated clutch of Fish and Game officers, burly, handsome, shorts-clad Fish and Game officers thigh-deep in the Henrys Fork, doing something unbearably manly with measuring sticks and water gauges, we basically forgot about the osprey and the sandhills and the...um, whatever else we were looking for in our purest-intentioned nature trip.

We had Fish and Game officers in their natural habitat. It was more than we could have hoped for.

In truth, they could have been something other than Fish and Game officers. One guy, I discovered from later species identification, worked for the Nature Conservancy. But that's really neither here nor there. They were men, Western men, with tanned faces and thick necks, smiling those wide Western smiles.

(Wait. I need to get a little sip of water here. Ooh, yes. That's better. My mouth went a little dry there for a second.)

There's something about a Western man. It's the sun and the mountains and the absence of social restriction, maybe. And a Western man with his shirt off standing against swift water is indescribable. We were enchanted. We slowed our canoes. (Actually Anne and I rammed our canoe accidentally into a bank of willows, but we'd like to forget that part.)

"Hi," one officer drawled. Anne and I, the less inhibited females in the party, had to brace ourselves against the thwarts of the canoes to keep from tipping out and swimming over for a closer look. "Nice day for a canoe. Where're you gals from?"

We stared at them for a second. They seemed oblivious to the effect they had on us. Anne tried to get to her camera, but I batted her hand away. Marla and I, the old married mothers of the group, recovered first.

"We're from Idaho," we managed to say.

Anne (2) chimed in, "I'm from Michigan."

"Hey, where in Michigan?" one strapping fellow called out.

"Lansing," she answered.

"Hey, I'm from.." and then he said the name of some town that was of intense geographical curiosity to Anne (2).

"How about you?" shouted another man, who was doing something exceedingly virile with a large dead trout and a hand-held scale.

He looked at Anne (1) curiously.

"Wyoming," Anne (1), who is actually from Pittsburgh, Pennsylvania, replied.

"Really?" he said. "Where in Wyoming?"

Anne (1) swore softly. "Where in Wyoming?" she hissed desperately at me.

"How should I know?" I whispered back. "I've never even been to Wyoming." Besides, my wedding ring had began to itch and burn on my finger and I had virtuously set about getting us disentangled from the willows.

Anne poked my lower spine with the tip of her paddle in retaliation and persevered without me. She is an attorney, so needless to say she can shoot the bull with the best of them. By the time the rest of our canoeing party arrived, she had that

Fish and Game officer, or whatever he was, convinced she was the next best thing to Annie Oakley. I stared at her in amazement as her voice took on a Teton-sized twang and her speech become punctuated with Western colloquialisms.

I've always attributed my love for Western men to the fact that I'm a Western woman. Never lived in the East. Never even been to the East. Maybe if I'd been born in Connecticut, I've always reasoned, I'd be attracted to stockbrokers with 30 suits in their closets and no idea how to skin a pig.

But as I watched The Annes, Easterners both, fall prey to those big-sky looks and that ah-shucks charm as easily as if they'd been raised in a shack on the sunny side of the Sawtooths, I realized it isn't just me. Anne (1) denied her family and her proud Pittsburgh heritage for a 5-minute flirt. Shameful.

I couldn't blame her, though. No telling what I would have said if my wedding ring hadn't started to itch and burn like that.

We saw sandhills that day, and osprey and moose. We even saw an eagle sitting on her nest. Seemed a little dull somehow.

After all, when you've seen Fish and Game officers in their natural habitat, lesser wildlife pales.

Love, loss at Dumping of the Beets

By Steve Crump

Name's Barnes. Jake Barnes. Fella named Hemingway conjured me up out of a bottle of rioja alta in a bar in Biarritz 70 years ago.

Swell fella, Hemingway. Swell as Americans go. Liked the fiesta of San Fermin at Pamplona. Came every year before the civil war. Good tipper.

Wrote about it in the Herald Tribune and the other foreigners started to come too. They liked foreigners in those days. Good tippers.

Sixty thousand show up every July for the Running of the Bulls. The locals run along with the bulls, and some of the foreigners too. The Americans get tossed, or take a bull's horn where they shouldn't. Great fun.

But not for me. Don't like tourists much, even the good tippers. They smell of cheap shaving lotion and take up all the good tables at Marona's cafe.

Came out to Idaho to talk to Hemingway about it, but found he's been long gone. Thought I'd go on to California to see Lady Ashley, who's married to a plumbing supply millionaire. On the way I pulled into a diner in a little burg called Paul.

It was the end of October, but warm. When I opened the door, it rang a little bell. At a table in the corner sat Brett.

"Hello, Jake," she said, rising to kiss my cheek and squeeze my hand. "It's a long time since the Hotel Montana."

"Hello, Brett," I replied, sinking into a vinyl-covered chair at a table covered with oilcloth. "Where's Mike?"

"Scotland for the golf," she said. "I got bored. Came here for the festival on a whim."

"What festival?" I asked, lighting her expensive Moroccan cigarette.

"The Festival of Sans Saccharine," she said. "To-morrow is the Dumping of the Beets."

I took a long pull on my Gaulois and blew smoke in her face.

"Never heard of it," I said.

"Started a few years ago," Brett said, holding back her ennui by tracing the tips of her manicured nails across the checkerboard pattern on the tablecloth. "There was an earthquake here that shook up a big sugar beet dump north of town. Millions of tons of beets rolled through the streets, sweeping a busload of German tourists away. A few got away. Word got around.

"So they built a bigger beet dump on an old cinder cone north of town. Holds 500 million tons. Every year, on the fourth Saturday of October, they roll all those beets down the hill and into town. Thousands of people, rich Europeans mostly, show up to try to outrun the beets. Some of them don't make it, Jake."

Brett looked away, biting her lower lip.

"Sounds like a bad business," I said, stubbing out my cigarette on a dirty pie plate and looking around for the waiter.

"But damned amusing," Brett said at last. "And I think I'm in love with a beet-dumper."

Brett looked ashen..

"I hold nothing against beet-dumpers," I said after a long silence. "As long as they're good tippers and don't write bad fiction."

Brett looked up, her eyes anxious.

"He's a good man, Jake. He runs a good beet dump. A strong beet dump."

"A clean, well-lighted beet dump?" I wanted to know. Brett turned away again.

The waiter came. I ordered absinthe, but they had none. Only flat Mountain Dew and maple syrup. I ordered the syrup.

"Jake, living in Europe has made you strange," Brett said at length, looking at the linoleum.

"And the war made me a bachelor," I said. "It doesn't matter. Let's have another drink."

When the waiter came back, I ordered another maple syrup, and then a third. I was beginning to feel tight.

"When do the beets roll?" I asked finally.

"At dawn," Brett said, taking my hand again. "Jake, promise me you won't run with the beets. I couldn't bear it if you were hurt."

"I'm no fool," I said, withdrawing my hand. "I'll find a cheap motel and a bottle of Fundador and watch it from the balcony."

Brett snapped her cigarette case shut and rose to leave.

"Arnold gets off work at the sugar factory at midnight," she said. "I must go."

I looked at the floor. "Brett," I said. "I've been a rotter."

Brett walked around the table and dropped slowly to her knees at my side.

"No, Jake," she said, looking full into my eyes for the first time since we`d met. "You're not a rotter. It's just the beets. It's been a warm October."

Brett smiled again, and then she was gone. I left a good tip.

To know me is to have seen me naked

By Suzanne Huxhold

CLOVER CREEK — I suppose if Scott ever dies I'll have to marry the other guy who works here. And not just for convenience sake, though there's something to be said for a short move.

True, the man's involved, and we don't love each other, except in the most platonic I-sign-his-checks-brother-sister sort of way.

Plus, I don't think his mother likes me. Not that that problem causes me any undue misery in my current marriage.

No, I'll have to marry the hired man because he's seen me. And when I say SEEN me, I mean that in the most biblical, standing naked at the ironing board kind of way. But don't force me to go into that. Neither the hired man nor I really want to relive the moment, frankly.

ANYWAY, my point is, marriage is a darn messy business. Nobody tells you that at first. It isn't anywhere in the marriage vows that I can remember.

"And will you, Suzanne, take this man, Scott, to be your lawful husband, even though sometimes he'll try to carry on conversations with you while he's using the bathroom and you'd rather he just close the door and shut up?"

Nope, that's the stuff you just have to learn the hard way. Like math. Marriage is the calculus class of the human race, only you every once in a while actually use the stuff you learn while you're married.

And if Scott died and I married our hired man, I wouldn't have to do remedial. We'd both go in knowing pretty much all there is to know about each other. We'd get 1600 on the marriage S.A.T.'s.

(Again, Scott wants to know why he has to die for me to get to do hypothetical things. As if he'll get out of this relationship any other way.)

I walked into the kitchen the other morning and there, at my table drinking coffee, was our hired man. At 6:30 a.m. And if I have to explain the implications of that to you, you are too young to be reading this column. He was waiting for Scott to get dressed.

I walked right in. Without a shred of shame. Walked in wearing my ratty bathrobe with the wine stain on the lapel from when I wore it to New Year's Eve (O.K., that's a whole other area we won't get into here) and the brown stuff of unknown origin on the shoulder I never did manage to bleach out.

And I had my slippers on. I can't really explain my slippers.

Just imagine wearing a sheep on your feet every morning for eight years, is all I'm going to say.

And my hair? It looked as if I'd slept with live bats.

So I sat down next to him and said, "Hey, does this look like ringworm to you?"

He dutifully studied my blotch for a minute. He thought it was probably just a rash.

"You eat strawberries?"

"Hmmm. Nope."

"Better put some Neosporin on that."

"You think it's contagious?"

"Hey, didn't you have a rash like that on your scalp one time?

Remember, they had to shave that patch on your head?"

See? Who but the man you're married to would have those kinds of intimate conversations with you in a dim kitchen at 6:30 a.m. and you in your ratty robe?

Who but the man you're married to has listened to you snore on an overnight cattle drive? Who but the man you're married to has ever seen you with the mascara you forgot to wash off the night before? Will talk to you about cattle prices when you haven't brushed your teeth yet? Will help you peg out the family underwear on the line if he's got nothing better to do?

Who but the man you're married to has seen you in a Mary Kay face mask and those little rubber toe separators you use when you paint your toenails once a year for the newspaper Christmas party?

If you live on a ranch chances are, your hired man. And I suspect neither of you are all that happy about it.

Love, like youth, is wasted on middle age

By Steve Crump

If, like me, you're an avid reader of Match Line, *The Times-News'* personal ads, then it may have struck you that romance doesn't age especially well.

In fact, it's easy to tell which supplicants of the heart are on the rebound after a lengthy first relationship. When they pitch woo, it's frequently low and outside:

"DWM seeks fun-loving, young-at-heart cowgirl, financially independent with no attachments, for intimate walks and possible whoopee. Have own pickup. No games."

"DWF with Rubenesque figure seeks fun-loving unattached, sensitive responsible man for companionship and possible relationship. Must like kids."

Now with all due respect to the second-time-around crowd, these ads couldn't get you arrested.

First off, they telegraph that you haven't been on the dating scene since "The Dating Game." And whenever somebody over 40 talks about being "fun-loving," he or she is telling you that they fall asleep by 9:30.

Then there's that troublesome word "responsible" – code for "my wife took the gold mine and left me the shaft." *Loser* alert! Financial institutions should be responsible; beaux should be able to cover the check.

"Companionship" means total celibacy and "possible relationship" means she's afraid of commitment. And "intimate walks?" Only from the truck into the bowling alley.

"No games" is a red flag for "guy with a spectacular history of getting dumped."

"Rubenesque?" Please. Why not just say, "Redux poster child?"

"Must like kids?" That means the woman has seven children between the ages of 13 and 21, six of whom are living in her two-bedroom apartment along with five of her grandchildren and her ex-husband.

"Cowboy?" Any guy who looks passable in jeans.

"Possible whoopee?" Take along an armed bodyguard.

"Financially independent" means "I'm not; hope *your* Visa's not maxed out."

"Unattached" means "hoping to avoid becoming a material witness in a bigamy case" and "sensitive" means you take your hat off at the table and your boots off in bed.

But "young-at-heart" is the worst symptom of all. Means you'll old enough to remember Loretta Young and you thought Robert "Father Knows Best" Young was cute.

Here's an appeal for full disclosure:

"DWM, without clue as to how to court, seeks like-minded female for on-the-job training. Bald spot expanding at slightly slower rate than gut, but like to dance if you don't mind badly. Not especially handsome, but not a jerk. You could do worse."

"DWF, single mother with way too many bills and way too little time, seeks law-abiding male citizen to talk about it. Don't scream at the top of my lungs and don't throw crockery because I don't have to. Interested in sexual relationship, but not right away. Mostly wanting nice guy who remembers the same presidents I do."

Isn't it romantic?

Headgames OK, but bring your ATM card

By Suzanne Huxhold

CLOVER CREEK — I was in Los Angeles last weekend. O.K., I don't know why either.

I picked up a copy of the "L.A. Reader" while I was there. The "Reader" is like a newspaper, but not really.

It gives you all kinds of information about movies and local rock bands and where to buy the perfect gelatto on Santa Monica Boulevard, which is apparently just crammed with gelatto, whatever that is.

It doesn't, though, give you any actual news. For that you have to read the "Los Angeles Times," and even then they don't give bean or soft wheat prices, so it's kind of a rip-off if you ask me.

Anyway, I'm thumbing through this "Reader" and I come to the personal ads. I can't resist.

I believe personal ads are the very mirror of a society. Personal ads are what anthropologists will someday read to find out about our different cultures. Look through any personal ad page in any city and see if you don't agree.

In New York, everyone wants someone to "experiment" with. In San Diego, "beach lovers" are big. In Boston, you got your "educated" type. In San Francisco? Well, in San Francisco it's a whole other deal, but you get my drift anyway.

In Los Angeles, apparently, the most important thing if you're looking for a mate is you don't want to play "head games."

Single females by the hundreds in L.A. are looking for singles males who don't want to play "head games." It's a No. 1 requirement, before, like, all their limbs and a good job. Single males also want women who don't play "head games," but it naturally comes after "athletic" (read skinny) and/or "pretty" (read implants).

Every single ad specified an aversion to "head games." Even the ones from "attached" (read married) males looking for "discreet" (read sleazy) relationships.

Uh, buddy, if you're already married and looking for a twinkie through the want ads, you've got plenty of nerve asking for a woman who won't play head games. I'm sure your WIFE would have something to say about head games. I'm sure your WIFE would like to play some games with your head. Like soccer. Or kick the can.

So this got me thinking. You know, "head games" would be at the bottom of my list if I were to place a personal ad.

Not that I would, as I am "attached." But say farmers and ranchers had their own "Reader." Nothing but agricultural news and bean and soft wheat prices and maybe a horse story every week.

(Hey!) We could have our own personal ads. Mine would read like this:

"Widowed ranch woman, (my husband wanted to know why he had to DIE instead of just divorce me for this column; I said he would never get out that easy), can operate backhoe, tractor and Brush Hog. Has own chickens, horse, D-8 Cat, shop, gas tanks, charge account at the co-op. Likes cowboy music and long walks through fields to check alfalfa for cut worms.

Looking for sensitive cowboy (Scott says this is an oxymoron; no sensitive man would ever BE a cowboy) with own tack. Must be good with children and able to buck 100-pound bales. Fencing experience preferred. "Head games" O.K."

And this is one I would answer. If I weren't "attached," ofcourse.

"Man with huge trust fund looking for athletic, pretty woman, 18-25, to help him spend millions."

OK, I'd have to fudge on some of that, and I know it sounds shallow, but really, who better than a widowed ranch woman to help a man spend millions?

What about the other stuff in my ad? Well, he could learn. He could learn.

Chapter 6
Bobwarr

Ladies and gentlemen: The Stones

By Steve Crump

A lot of guys go to summertime get-togethers to pick up women.

I went to the Gooding Basque Festival in Gooding Saturday to pick up rocks. OK, not me personally — but I watched. You just gotta admire fellas who can heft a 300-pound boulder and walk away without walking away funny.

Weightlifting and weight-carrying are traditional Basque sports, of course, which helps explain why the Basques fit in so well out here in Rockchuck Rendezvous. I don't care what the Idaho Potato Commission says:

Our leading commodity here is really basalt.

It grows in abundance from Montpelier to Payette, great brown-black lesions that stand out like warts on the nose of a debutante. Idahoans have been working 125 years, man and boy, to cleave, blast, pry, drag, coax, wheedle and motivate rocks out of the paths of our tractors, but they always come back tougher and meaner and a little heavier than before.

All true Idahoans are rock-pickers, whether they were to the homestead born or not. Hired out for a summer's pipe-moving, city lads used to spend three mornings out of five standing hip-deep in mud trying to move a boulder the size of Rush Limbaugh, only smarter.

The fella named Howard worked the neighboring farm where I grew up in Arbon Valley, and Howard made it his life's work to collect all the lava rocks on his place and to pile them up in a big stack next to the grain silo.

He was 50 or so by the time I came along, and the rock pile was already taller than the silo.

Then a funny thing happened: Howard had a heart attack and was laid up for a while, and his boys stopped picking rocks in the fields. Little by little, that rock pile began to shrink. Howard's wife, Della, swore that when nobody was looking, them rocks crawled back into the fields.

I'm sure you'd find them there today, along with their pebbles and grand-pebbles.

That's not the case in Basque Country, I suspect: The Pyrennees are only so tall, and when they're gone, they're gone.

Idaho, by contrast, will be about 30,000 feet above sea level by then, for out here on the Snake River Plain, the rocks are winning.

That's because rock-picking is a dying art. For starters, it's backbreaking work, which is something that today's air-conditioned tractor crowd eschews whenever possible.

The avoidance of picking up rocks has even become something of a social ritual, as you'll see if you drive down any farm-to-market road in the Magic Valley.

You'll soon come upon three pickups pulled off the road and three farmers, standing in a semi-circle in the field with their hands in their jeans pockets.

Before them will be a 3-foot-by-2-foot Gibraltar wannabe, resting in 8 inches of volcanic soil.

"Well, Jeb, you could blast 'er," Farmer 1 says.

"Yep, but I'd have to go to town for caps."

"Well, you could put a line on 'er and haul 'er out," Farmer 2 says.

"Yep, but my hitch is broke."

"Well, you could leave 'er here and buy a new plow first time you run over 'er."

"Well, hell. Need a new plow anyways."

The Rev. Derwand McEldowney, late pastor of the First Southern Baptist Church of Soda Springs, would have been appalled.

Rev. Mac, who also dry-farmed 640 acres of winter wheat, was pretty sure that everyone he'd ever met was tap-dancing on the trap door to hell. And he considered slackers — defined as anyone who couldn't strike a match on his calloused hands — to be the shock troops of perdition.

One night, Rev. Mac caught his son, Jeff, and two other boys, Dave Miller and Arnie Scroggins, drinkin' sloe gin in a old duck blind down by the Bear River. Before dawn, he put them to work on The Point.

The Point was an massive outcrop of basalt that jutted out into one of Fred Larsen's fields. Fitful attempts had been made over the years to dislodge it with dynamite, leaving a jagged column about 20 feet high and 50 feet wide, surrounded by broken rock.

The Rev. Mac handed each of the boys a new shovel and said simply, "Move it."

The Point is still there, but it was well taken. The boys turned out OK after all.

Jeff is now mayor of Soda Springs and Arnie is a Mormon bishop. Only Dave failed his lesson.

He lives in Pocatello now, where he runs a medical-supply business.

Yes, they carry trusses.

Looking for a man? Get a job at Ready Mix

By Suzanne Huxhold

CLOVER CREEK — I went to the Laundromat the other day and picked up a magazine called "Working Woman," if you can believe that. After I scanned the section on colored pantyhose (apparently, taupe means you're serious about your job, red is kicky but shouldn't be worn with a nurse's uniform and eggshell makes your ankles look fat, as if we didn't already know that), and the article about inter-office sex (I asked the guy in my office what he thinks about sex in the workplace and he said it was the only reasonable alternative considering the cats were always in our bed), I read a riveting item about the scarcity of decent men. Seems working women are having a hard time finding any.

Pish-posh.

They just have the wrong jobs. You want a man, you learn to pour concrete, is my advice. You'll be hip-deep in men. You won't be able to swing a dead dog without hitting one (and what woman hasn't had THAT fantasy).

Concrete is a decent-man magnet; a more powerful seducer than football or Baywatch or double nacho night down at The Oasis. I've seen men so riveted by the sight of concrete you'd think it had been poured around their shoes and left to set. All a woman in this day and age has to do is learn to pour concrete and men will come racing her way. It's true. I've always known this secret to catching a man with concrete, you know. My father was a construction worker, my brothers and uncles the same. If there was concrete being poured anywhere in the state, it seemed to me when I was growing up, there they would be, standing around a hole in the ground, staring reverently at the oozing stuff as it globbed from the back of a grinding barrel like the blue cheese dressing you forgot at the back of the refrigerator for eight months.

They'd discuss slump and yardage and gravel size as if, in the right combination, they'd be the formula for happiness and eternal life. If they were especially lucky, they would be allowed to float and smooth it and scratch their initials in it with a little stick; if they were not, they would only wait and watch it dry.

I always thought the men in my family were damn stupid, frankly.

But I know now it isn't just them. According to the half-dozen or so men who work around here, stringing fence lines is near torture, moving cows is a chore, feeding hay is so boring it's not to be believed. But pour cement in a hole - and even a bag in a wheelbarrow is sufficient to catch their attention - and they'll be running over each other to participate.

There's a small project ongoing at my son's school, a building being brought in to house the first grade. None of the men in town paid much attention to the project at first, not until someone happened to notice there were concrete forms being built.

Concrete forms. The duck call, the elk scent, the very Playboy Channel of the decent man. Men come from all over now, stopping at school on their way to and from the potato fields and sale yards and feedlots to watch the process. Ignoring the children that scream and soccer around them on the playground, they watch in little decent-man knots, intent only on every move the concrete guys make. They have to stuff their hands in their pockets to keep from twitching with the desire for a little trowel.

"You know what they say about people who watch cement dry," I joked to a clump of the concrete sentries one morning as they huddled over the hole in the ground. They just stared at me.

A divorced friend of mine called me yesterday. She'd read the "Working Woman" article, too.

"It's TRUE," she wailed. "All the men at my job are married, all the men at the bars are scum and everybody else still lives with their parents."

"Eeew," I said compassionately. "Not really with their parents?"

"Yes. With their parents."

I knew what had to be said. I owed it to her. "You know what you should do?," I said.

"What?"

"Learn to pour concrete."

"WHAT? But what about my job, my education, my business wardrobe?"

"What good is a $50,000 a year career, a master's degree and a closetful of Donna Karan when the only guy you can get to date you doesn't want to go out Saturday night because his mom makes fudge during "Dr. Quinn, Medicine Woman?," I countered reasonably.

I was on the phone with her for three hours, but I think I finally talked her into it. She was heading to the store for supplies right after we hung up. I told her I thought pantyhose in a nice shade of grey would be best.

Who needs bridges from Madison County?

By Steve Crump

"The Culverts of Gooding County"

Fresca Jondyce sat in the porch swing and shelled chickpeas, nursing an iced coffee and a grudge.

She had gone to a roadhouse in Bliss the night before with her husband, Emmett, who was a clown.

Emmett, the sole heir of 7,000 acres of prime Gooding County stinkweed, had taken to bulbous noses and floppy, oversized shoes about the same time the mule kicked him in the head. Now he had taken to cruising Wendell in his full clown regalia, ogling the attendants at the new coin-operated truck wash.

So Fresca went along to keep an eye on him. Her left one, usually.

Emmett started out the evening at Chez Malad with boilermakers, graduated to aspirin and Coke, and ended up getting bounced for throwing bucketfuls of confetti at the bartender.

Fresca sat there sullenly, taking it all in, the buzz from the caffeine from the ceaseless stream of Mountain Dews making her pine for home, for far-off Chubbuck.

Yes, she was a hot-blooded, sensual Eastern Idaho Democrat in a community of Alf Landon Republicans, and the locals viewed her with a mix of suspicion and bemusement.

After all, she'd married Emmett only because she was extremely near-sighted and not wearing her glasses when he proposed, and that she was accepted at all had mostly to do with the fact that she kept Emmett from attending children's birthday parties to which he had not been invited.

Most of Fresca's friends had drifted away after that fateful date in '63 when she had stood up in a council meeting and boldly advocated fluoridation, but her life was full of whoopie cushions and rubber chickens, and her two teen-age children, Chuckles and Claribel.

But today, with Emmett, Chuckles and Claribel away in Boise for a few days at a big Shiners' parade, she was ready to turn a page in her life. And on that page, she found Rupert Champale.

"How do', ma'am?" Rupert said, stepping out of his battered GMC pickup just before it ran into the back of the chicken coop. "I'm looking of the culverts of Gooding County."

Fresca arched her back and stiffened. The culverts were her secret passion.

As a child, she had scampered over concrete pipe and dreamed of one day taming the mighty Portneuf River. After she graduated summa cum laude from Cal Tech and the choice came between building the English Channel tunnel and going to work for the highway district, Fresca never hesitated a moment.

But Emmett didn't like Fresca working, so she quit and stayed home, mopping the linoleum in the kitchen and inventing the Space Shuttle in her spare time.

And now Rupert Champale was here to take pictures of the culverts of Gooding County for Hearth & Tractor magazine.

"I'll take you there," Fresca said, her heart racing on courses unimagined by her neighbors.

Or by Rupert Champale himself. At 52, he had shot every antique tractor worthy of the name, and his avant-garde photo essay on center pivots had been exhibited twice at the Metropolitan Museum of Modern Art.

Still, he was unprepared for what he found when his battered truck rolled to a stop along Billingsley Creek.

It was a 12-foot culvert of gravity flow construction, capable of handling a four-digit cubic-feet-per-second flow. Champale`s hands trembled visibly as he raised his Nikon.

"We are part of a larger whole," Fresca whispered, steadying his hand with a touch. "Prisoner of love, wind beneath my wings, town without pity."

"Cara mia," he sighed, and they kissed, the angry waters of Billingsley Creek roiling about their feet.

A year later, Champale's photographs of the culverts appeared in Hearth & Tractor, making him an international sensation and real popular in Gooding County. By then, Chuckles and Claribel had joined the Ringling Brothers, Barnum & Bailey Circus and Emmett had been elected to the Legislature.

But Rupert was dead, the victim of a corn-threshing accident while photographing new silage techniques in the Ubangi, and soon thereafter, Fresca herself disappeared — vanished into the high August sky on a day very much like the day she met Champale.

But there are those who say, in the stillness of the morning or at twilight, that you can still hear the wail of the lovers, locked in an eternal embrace below the Billingsley Creek culvert.

"I can't go away with you, my love!" Fresca's spirit cries. "I'm standing in two feet of mud!"

Granny

By Suzanne Huxhold

CLOVER CREEK — I had lunch at the cafe in town the other day. A friend of mine was there with her family and her embryo transfer vet. They obviously hang out together. Something about those embryo transfers that just draw people closer, I guess.

Anyway, when my friend introduced me to her grandmother, the darling old gal looked me dead in the eye and said, "I read your column on mice. I had 'em a whole year before you did."

It was a challenge, pure and simple. I know one when I see one. Oh, she may have had the twinkly granny grin and the white hair and the ribbon-candy look about her, but she was calling me out.

"I've got voles now," I said, never one to ignore a dare.

"Millions of them. They're tearing up my lawn."

She made that pishing sound through her pursed lips old ladies make when they think you're a fool.

"Voles?" She waved a frail hand in the air, dismissing me. "We had voles last year. And they didn't just eat my lawn. They ate my grapes and my trees."

"My voles ate the roots of all my raspberries."

She raised a snowy brow. "Just the roots?"

Oh, she was smug.

"And the foundation of my house," I added. "We're sitting right on dirt now." Which was a total lie.

"You should get Mormon crickets sometime," the embryo transfer vet ventured. "They're what's bad."

I glared at him. "Stay out of this," I hissed. "This is between me and granny here."

But she was delighted.

"Oh, I've had Mormon crickets," she said loftily. "Mormon crickets are about the worst thing you can get."

Damn. We've never had Mormon crickets. I'm not even sure what Mormon crickets are. Grasshoppers with nice big families, perhaps.

"We've had army worms," I said.

She rolled her eyes. "Army worms?"

Vaguely ashamed, I pressed on.

"And drought. Seven years of drought."

She smiled. "Ever heard of the dust bowl?"

"And floods," I said, feeling a little desperate now. "Twice our whole place has been flooded. We had to canoe out to the school bus stop."

"A school bus comes out to where you are, dear? You must practically live in town."

Oh. Low blow.

"And besides," she continued in that sweet, creaky little voice, "floods come and go. What's your annual snowfall?"

Hell. Our annual snowfall is crappy. Hardly even biblical. I sized her up. She was a Nevadan, from down around the Rubies. What could they get there in a winter? six feet?

"Oh, we get 12, 15 feet most winters," I lied, having forgotten for the moment her granddaughter lives 10 miles from me and would know what a whopper that was. "Sometimes the plow comes out and you can't even see it over the snow."

"You have a plow that comes to your house?" She made a show of straightening the napkin in her lap. "My, that must be nice."

"Lightning fires!" I shouted.

She just shook her head.

"Sustained winds of 52 miles an hour!"

"A mere breeze," she said.

"Thirty-six below zero on New Year's Day!"

"Ooh. Sweater weather."

"Rattlesnakes!"

She dug a handful of rattles out of her purse and tossed them on the table. "Killed these on my back stoop this morning."

I was defeated. I knew it. My plagues and problems were nothing compared to hers. And I've always been so proud of them.

"We have cattle," I mumbled, a last ditch effort.

She sat up, alert, like a boxer in a ring. "Dairy or beef?"

"Beef."

"All right," she said, throwing her napkin next to the rattles. "You win."

CHAPTER 7
KINFOLKS

The future, in a photo finish

By Steve Crump

The future wears its pants pulled clear up under its armpits and its teeth in its hip pocket, spits tobacco juice into the hydrangeas, and complains about the government full time.

I discovered that in a family photo album, of course.

Therein can be found photographs of my grandparents, parents, aunts and uncles at various pit stops along the road to their horizons. And every dang one of them looks like me.

They're all there: Great-Uncle Clyde, who had a cowlick that would have shaded a small flock of sheep; Grandpa, whose brow was so craggy that he made Lincoln look like Twiggy — even Great-Aunt Patience, who wasn't.

The dates are on some pictures, which makes it easy to do the math. Some of the photos were taken when these folks were 50, 45, — dare I say it? — 40 years old.

That can only mean one thing: Any day now I'm going to have one eyebrow, two teeth and three chins.

You could look it up: When it comes to sparing progeny from the signs of their fathers, genetics has no sense of humor.

And my gene pool is as clear as a cracked mirror. All you need do is track the photographs of my forebears from the cradle on, and it's obvious what I have to look forward to.

Same chubby cheeks, same bugle beak, same jug ears. The unmistakable conclusion is that I'm not merely about to look like my kinfolk; I'm about to become them.

That sunk in recently as I came across one particular picture of my grandparents, my mother and her seven brothers and sisters, along with my grandpa's brother and sister-in-law and their kids.

The photo was taken in the Cedar Draw area near Buhl shortly after World War I. The photo quality isn't the best, but the faces are easy to discern.

Each of them looks oddly familiar, like a snippet of a glance you catch while shaving in front of a fogged-up mirror. Here's a fragment of chin, there's an angle of nostril, over there the arch of an eyebrow.

It's sort of like a police artist must feel when he's compiling a composite drawing; elements of each of those faces make up the whole, which is to say that if you put all of those faces together, you'd have mine.

Darn scary, I must say.

It's probably only a matter of months before I'll have to get my hair cut with a bowl on my head, buy a collection of bib overalls, and back over my Stetson with the truck.

By then, I'll have collected "Walter Brennan's Favorite Hits" and have 200 hours of "Hee Haw" on tape, spent my summer vacation in Chautauqua camp, and bought new high-top dress shoes from the Monkey Ward catalog.

I'll be combing my hair with a fork and staying up until 8:30 on Tuesdays just to listen to Lowell Thomas on the Philco radio.

And if my wife kicks me out of the house, I'll just go sleep in the Ford truck body behind the pump.

All of which smacks of moral determinism, the notion that you ought to look like what you become. Could it be that Dorian Gray wasn't a victim of his own evil, but just bad genes?

And yet I don't remember all of these relatives who grimace back at me from the pages of the family album as being a particularly homely lot.

They were just ordinary folks, hard-working Idahoans who kept their fears to themselves and their hopes in an old coffee can buried in the back yard.

Aunt Martha gossiped a little too much and Great-Uncle Tim used to get drunk at the rodeo, but on the whole, they were good people, deserving of a better fate than American gothic.

So I guess if hillbillyhood, or at least jiggling jowls and an ability to balance a cane chair on two legs for hours on end, is my fate, then I'll go gently into that good night.

There are worse fates in life, after all, than plainness. Bill Monroe, the bluegrass king, used to sing about a man who'd done too much honky-tonkin' to be worthy of his backcountry mama, a woman of such righteous simplicity that she could forgive even what she could not understand.

But please, God, just don't make me worthy of Aunt Beulah.

Really, I could kill these people

By Suzanne Huxhold

CLOVER CREEK — I was pretty sure I was going to die of cabin fever last week, until I read about the poor Clear Springs Trout guys who were stuck in the cab of their 18-wheeler during a three day snowstorm in New York.

That would give a whole new meaning to cabin fever, wouldn't it? I wonder which one of them is still alive.'Cause, of course, one of them had to kill the other, don't you think? I've entertained thoughts in that direction myself during this past week of storms, and I have a bathroom here. And Pop Tarts.

Which is not to say I don't like raw trout as much as the next guy. But if I had to LIVE on it, PLUS wear the same underwear, PLUS be in New York for three days without seeing "Cats," I think I'd go mental.

Not that we didn't have our own problems during that snowstorm. For instance, I almost murdered my entire family the day they called off school in Bliss.

At 7 o'clock that morning the wind was howling and the snow was coming down horizontally and the drifts had completely covered my car, which I'm pretty sure I won't see again until Easter. Even the snow plow had gotten stuck at 4:30 that morning, for half an hour, right in front of my bedroom. Have you ever tried to sleep while listening to a snow plow guy swearing at the top of his lungs? It's not easy.

So I was thrilled when my neighbor phoned and told me school was canceled. My son and I fixed hot chocolate and read books and made cookies. I felt like Louisa May Alcott.

But by 11 a.m., when his energy time bomb exploded - "Mom, Mom, want to play Monopoly? Mom, Mom, it looks cold out there."

"Should I let in all the cats? Mom, Mom, look, I found Dad's screwdrivers in the laundry room. Can I take apart the dishwasher!"

"Mom, Mom, how does electricity work?" - I was acting more like Stephen King.

"Stop talking to me!" I shrieked, grabbing at my hair like a junkie on a bad trip.

I was ready to bundle him up real tight and let him take his chances with Mother Nature. How bad could an 11- mile walk to school in a blizzard be anyway, I was thinking." He has a hat.

Luckily, my husband was much calmer.

"I'm killing your damn chickens," he announced when he came in from feeding. "Heat up the pot, baby, we're having DUMPLINS' tonight! And the cats. I'm also killing the cats. I'll be linin' my boots with CAT FUR tonight, baby!"

(Our son let out an ear-piercing wail at this. He's usually at school during his father's insane morning outbursts. He doesn't know how to wait them out like I do.)

The chickens, apparently, decided to lay a pile of eggs in the engine of his tractor sometime last fall. I know it must have been last fall, because that was the last time he used the tractor.

And because when he started the tractor so he could plow the driveway, and left it running until the heat from the engine popped all those little brown eggs - which filled the valley with a stench you would not BELIEVE - it left little room for doubt.

And the cats, of course, won't leave the porch while it's snowing to do their business. So their business was done that morning on the porch. And when Scott took off his boots to come inside, he happened upon their business in his socks.

And then there was the food situation. We're a long way from town. When it snows, it's even longer.

So naturally we turned into ravenous wolves. We ate everything in sight in some demented attempt to store up fat, like we need to, and then stood in front of the freezer and screamed, "Oh, God, there's nothing to EAT! We'll STARVE out here!"

At which point Scott went out and tried to kill a chicken.

Frankly, he'll use any excuse.

Sure, we weren't in the cab of a semi-truck when all this stuff happened, but I'll bet those Clear Springs guys didn't have any their families with them, either. See, it all evens out.

A little clean fun while Mom's away

By Steve Crump

I need a woman's opinion.

You see, my wife has been in Pocatello visiting her mother, and she's due back this week. Due to a series of unfortunate although entirely innocent events, the house is – What's the word? – a bit of a picker-upper.

OK, it looks like a cowboy bar after two-for-one beer night.

Truth is, the scope of the damage exceeds the time that my two teen-age sons and myself have left to clean it up. We're going to have to prioritize here.

My question for you, ladies, is which of following outrages is most likely to land me in a full-body cast. Please rate them on a scale of 1 to 5 – 1 being exile to the Dry Tortugas and 5 being getting my personal affairs in order.

Incident 1: Last Sunday morning, my son was preparing breakfast, and is his wont, he made scrambled eggs.

Now scrambling eggs normally requires extensive stirring, but all of the spoons being dirty – we last did the dishes on the day Mom left – he thought it might be fun to take the skillet and toss the eggs, flapjack-style. Then he, well, kind of overshot the mark.

The eggs splattered over a four-square-foot area of the kitchen ceiling.

About two feet short of being able to reach the area of impact, my son grabbed the dog, lifted her up to the ceiling and encouraged her to consume the evidence. She demurred.

Quickly assessing the situation, and being late for the kick-off, I assured the lad that, being fluffy, the eggs would "probably fall off by themselves."

Physics lesson here: They don't.

In fact, they're glued to the ceiling like the front-porch railing to your tongue on a 40-below morning.

How much trouble am I in?

(Circle one)

1. Keep your head down.

2. Egg-cellent prospect for compound fractures.

3. Consider blaming the dog.

4. You're headed for the couch, and the couch is headed for the garage.

5. Hide in the neighbor's basement.

Incident 2: My wife has – OK, had – this expensive lead-crystal thingamabob that sat on top of the TV set. When my son went short and I threw long, it sort of got broke.

So, for that matter, did the TV.

How much trouble am I in?

(Circle one)

1. Go to the florist, have him set aside one dozen red roses, then send every other rose in the shop to your wife.

2. Good chance you'll be watching the Super Bowl at the Do-Drop-Inn.

3. " ... then that darned ceiling fan blew apart and missed us by *that much.*"

4. Four words: "But I can explain!"

5. Fourth-and-30.

Incident 3: We ran out of soap for the dishwasher, so I figured, "What the heck? I'll substitute this dishwashing liquid. If I use the whole bottle, it'll get the barbecue grill really clean."

When I came back inside the house 30 minutes later, the entire kitchen was engulfed in suds. We still can't find the dog, much less the sink.

How much trouble am I in?

(Circle one)

1. You're about to get washed right out of her hair.

2. Get out the box fan, blow the suds into the living room and explain that you're shampooing the carpet.

3. Practice saying, with a straight face, "Oh, nothing much, dear, how was your trip?"

4. "What do you mean, where's the kitchen?"

5. "Honey, we've been named a Mr. Clean poster family!"

Mail your votes to You're Toast! c/o *The Times-News*, P.O. Box 548, Twin Falls 83301, or fax them to me at 734-5538. Next Sunday, I'll let you know the results.

Say, does your doghouse sleep two?

City mouse, country coyote

By Suzanne Huxhold

CLOVER CREEK — My little sister and I have always been very close.

The Little Girls, the family called us when we were children, never just Suzi, or just Teresa. The Little Girls, because they could no more separate us in name than they could in body or spirit. I can't remember a time we didn't share our bedroom, our dolls, our secrets. We were best buddies, constant companions, true confidants.

Which would lead one to believe Teresa and I are much alike.

Well, one would be wrong. Big time.

Teresa and I, The Little Girls, are a study in opposites. I look like a Viking, she looks like a gypsy. She was president of the business club in high school, I was best actress. She's an executive in a multi-national corporation in the third-largest city in the United States, I usually end my work day by scraping manure off my clothes. The last pair of shoes I bought cost $9, because I splurged and got leather uppers. The last pair of shoes she bought cost $87, because she got them at a half-price Easter sale.

She called me today from her suburban home near Chicago.

"What did you do today?" she asked me.

"I helped Scott move heifers over from the Hess field to the upper creek field." She doesn't know where either of those places are. "What did you do?"

"Oh, you know. Business stuff." I wouldn't understand, she thinks, and is dead right. "I found a mouse behind the dryer in the garage when I came home from work."

"Yuck. While we were riding through the sinks, a coyote came out and started yipping at us."

"Sounds scary. I was afraid to pick it up. You know, hanta virus or whatever."

"Yeah. Four people in Idaho have died from that, I think. I've never seen anything quite like it. We had all the dogs with us. She must have been trying to draw them in closer to her den."

"Cool. Anyway, I didn't know what to with the thing. I had to call the city about it."

"She got so close I could see the whiskers on her nose. What do you mean you had to call the city?"

"I had to call the city. About the mouse. I couldn't just leave it there. It made the whole garage smell. Were your horses spooked?"

"A little. Apache kept tossing his head. It was just a mouse, right? Why didn't you just pick it up and throw it in the trash?"

"I wasn't sure if you're allowed to throw dead animals in the trash in Barrington. They may have had an ordinance or something."

"It wasn't a doberman." She had my full attention now. "You could've launched it onto the mayor's front porch and he wouldn't have noticed."

"I knew you would laugh."

I tried not to. "Why didn't you have Randy take care of it?"

"He's out of town. I thought about burying it, but we don't have a shovel."

"You don't have a..? Jeez, Tedda, what kind of homeowner are you?"

"The kind with a gardener, thank you very much. Did you shoot her?"

Sisters, I thought. I knew just where the conversation had picked back up. "God, no. She was so pretty. She followed us for a half mile, then followed us again when we came back through. What did you finally do with it?"

"Well, the city guy called back after checking with his supervisor and told me I could just put it in the trash at the curb. They'd pick it up and dispose of it for me."

"Big of them," I snorted.

"It was gross. I used the dustpan. I had to soak in the tub for an hour just to get the smell off."

I looked down at my jeans. God knew what that brown stuff was.

I picked at it with my thumbnail. "Tedda, does it ever occur to you one of us just had to be adopted?"

"Well, I just told our sister the same story, and she said she would have called an exterminator to get any old mouse out from behind her dryer, so it must have been you. Do you think she had pups?"

"Our sister?"

"Oh, you're a riot. Talk to you tomorrow."

"Know what, Tedda? I can't wait."

In search of the Grail

By Suzanne Huxhold

CLOVER CREEK — I've always wanted my own apartment. Well, not always. Just since I got married.

Oh, romantic devil, I can hear you say. Well, I AM. I am romantic. I'm crazy about my husband. I simply want separate living quarters.

I grew up with six brothers and sisters. Privacy was the Grail in our house; much sought after, never found. I never had my own room, never got into the bathroom first, never shaved my legs with a new razor, never did laundry without someone else's underwear mixed in with mine. You can see how that would wear on a person. I think underwear washing should be a semi-private thing. Like crying. Or a prostate exam.

And I went from washing my underwear with people I'd known all my life to washing them with some stranger's whom I'd got myself tied to, in a rash moment, for all eternity.

I tried to set privacy boundaries with the new person, too, but he wouldn't have it. I never knew when he was going to waltz in on my personal life and demand marital attention or a sandwich or something. I had less privacy with the one stranger than I ever had with six siblings.

Then, the child was born. Boundaries? TRY to set boundaries with something that requires you to unbutton your blouse every time it wants a drink. I was a 24-hour biker bar at a Hell's Angels convention for all intents and purposes.

It's no better now that he's older. Now there are two completely unself-conscious males in my house who think nothing of perching on the toilet seat to chat while I stand naked in front of the bathroom mirror and try to see if that's a mustache on my upper lip or just some cruel trick of the light that's making me look like my grandmother, who used to have her own electric razor. I tried locking them out, but they pick the locks. No, I'm not joking.

The apartment thing became a bit of an obsession, as it naturally would.

One more intrusion on my privacy, I would threaten on occasion, and I'm WALKING! They'd just stare at me, blankly, then ask me if I was going to do a load of laundry, as they were both out of underwear.

Then, last week, my husband went fishing with my dad and my brother and my uncle. He was gone seven days.

Ah, sweet privacy! With the child in school all day, I would wallow in it. No one popping in for a snack, or what have you, at odd moments. I'd be alone, blissfully, blessedly alone, for the first time in my life.

The first day I fixed myself a pot of tea, read two entire romance novels and ate cereal for lunch.

The second day I fixed myself another pot of tea, cleaned the bathroom, then wandered from window to window, checking locks.

The third day I skipped the tea, locked myself in the house and began muttering to myself like a crazy old man.

"Need to fix the doghouse, need to fix the doghouse. Doghouse needs fixing. Should get to it."

The nights were worse. I slept with a gun. Can you believe that? I read the owner's manual that first night at 3 o'clock in the morning because I was up anyway, searching under all the furniture, so I was certainly prepared for any eventuality, but a gun? How stupid. Not once in 10 years on this ranch have I ever felt the slightest threat, but put my husband on a fishing boat in the Sea of Cortez and I turn into Xena, Warrior Princess, only without the breastplates.

One night, when my son went off to sleep at a friend's, I woke up to the sound of the dogs barking across the creek.

"Can you hear those damn dogs?" I muttered, knowing full well I was alone. Completely alone, just like I'd dreamed. "Damn dogs.

Damn dogs. Always barking. Bark, bark, bark." I was tempted, just for a second, to bark myself, to howl, just to hear the sound of it in that empty bedroom.

I don't want my own apartment anymore. I see now with perfect clarity it would drive me insane. I'll mix my underwear with these guys' any day of the week.

Chapter 8
Mud flaps and 8-tracks

Lost in Twin? That's neither here nor there

By Steve Crump

A former co-worker from Idaho Falls who had never been to Twin Falls before paid me a visit last week.

Tried to, anyway.

Jerry called from the Traveler's Oasis and asked for directions.

"It's simple," I said. "I work at the corner of Third and Second."

"Come again?"

"Third Street West and Second Avenue West."

Dead silence.

"... I flunked algebra in high school," Jerry began. "But even I know Second and Third streets don't run into each other."

"Not Second and Third *streets*," I said with exaggerated patience. "Third Street and Second *Avenue*. They're not the same things."

"If you say so," he said. "See you in a few minutes."

An hour passed, and the phone rang again.

"I'm at the corner of Second and Third, and you're not," Jerry said.

"That doesn't tell me anything," I said. "What's the nearest building?"

"A white house," he said.

"Could you be just a shade more specific ..." I wondered.

"More specific?" he exploded. "How many intersections of Second and Third can there be?"

"In Twin Falls? Eight," I said calmly.

"Eight!" he screamed. "What kind of place is this?"

"Twin Falls," I said soothingly. "The land of the lost."

"Who in his right mind would create eight intersections with the same name?"

"Some with a warped sense of humor who'd been camped in the sagebrush for too long," I said. "Listen, it's really very simple. Think of a baseball diamond, and draw two imaginary lines, one from home plate to second base and one from first base to third base."

"So where does that put me?" Jerry asked.

"I have no idea."

"What?"

"Because I don't know *which* intersection of Second and Third you're at," I tried to explain.

Jerry slammed down the phone and disappeared for a full minute.

"The sign says Third Street North and Second Avenue North," he huffed upon returning.

"OK, you're where the shortstop would be lying if he tripped over the pitcher's mound," I said.

"What?"

"You're two blocks east of where you're supposed to be."

"How can I be two blocks east of where I'm supposed to be if I'm at Second and Third North and you're at Second and Third West? Since when is north east of west."

"Since Ira Perrine said so in 1904."

"I'm getting dizzy," Jerry whined.

"Look, all you have to do is head two blocks east and you're here."

"No, I won't."

"Why not?"

"Because if I travel two blocks east I'll cross Second Avenue North, then First Avenue North and end up at the corner of Main Avenue and Third Street West."

"There is no First Avenue North. There are no First Avenues anywhere in Twin Falls."

"How can you have a set of numbered streets and no First Avenue?"

"Because Main Avenue is First Avenue."

"You just said there was no First Avenue."

"There isn't. Listen, just find Shoshone Street."

"Wouldn't it be easier if I just found First Street."

"There is no First Street. Shoshone Street is First Street."

"Look, I'm sorry, I'm just going to find Shoshone Street, head due north and go home."

"If you head due north, you'll end up in somebody's garage."

"What?"

"Because downtown Twin Falls is laid out at a 45-degree angle. There's no such thing as due north."

"I'm never going to get out of here, am I?"

He finally did, or course. I went to lunch, and when came back there was a phone message from him saying he had found his way to Fourth Avenue West and was headed for the freeway and Idaho Falls.

I wondered if he realized his mistake before he got to Wendell.

Pals

By Suzanne Huxhold

CLOVER CREEK — Ranching is crammed with men. You can't chuck a rock in this business without hitting a man, not that I haven't tried.

They run all the saleyards and the feedlots, most of the ranches, even the trucking companies. You see 'em gossiping in the coffee shops, drinking beer in the cowboy bars, sitting on selenium blocks in the co-op, jawing about calf prices and the way the world's going to hell in a hand-basket.

They're everywhere, men, and though most of the time I relish their company, sometimes I swear if I hear another old coot snorting into a dirty red bandanna while I'm trying to eat, or listen to another long rant about Hillary Clinton or watch another guy scratch at stuff I don't even want to THINK about, I'm going to scream.

That's why my women friends mean to me what they do. They are more than confidants and comrades; they are refuge from testosterone and seed company caps and the idea that if you talk about the weather often enough, you might be able to CHANGE IT.

My pal Terrell and I went to Elko a couple weeks ago for the Cowboy Poetry Gathering. We met at her house in Wendell at 5 o'clock in the morning for the four-hour drive. Terrell is a most competent woman; she can handle a skittery horse and manage a complicated career and relate to her two teenage sons like nobody's business. But drive, Terrell cannot. At least, not particularly well.

I nagged her about it until we got to Wells, at which point she turned to me and said, "I'll bet you cook from the dining room, too." This, I thought, was the funniest thing I ever heard. I snickered and coughed while she climbed to the back of the camper- van to let me drive the rest of the way. She took over again once we got to Elko, but after she drove into the same private driveway three times on the way to the convention center, and I had to undo the top button on my pants because I was laughing so hard, I confiscated the keys. This set the tone for the whole weekend.

Terrell would make her way through the thick knots of poetry lovers by turning to me and saying, "Walk this way," and I would respond, on cue, "Well, if I could walk THAT way..." and we would snort and giggle until people started to stare. Her font of lame jokes and bad puns was endless, as was my ability to be absurdly amused by them. By the time we ate dinner at a Basque restaurant Friday night, we were giddy with freedom and cowboy poetry and everything seemed funnier by a million miles than it would have if we had been with our husbands. When our waitress came around to ask if she could bring us anything else, we looked at the two heaping plates of lamb, the french fries, the salad, the soup and spaghetti and beans and bread and peas and I slammed a fist on the table and said "BRING MORE FOOD," all three of us, the waitress and Terrell and I, had to lean against each other to keep from falling into a limp heap on the floor.

(We left the woman a $10 tip. Anyone who thought we were as funny as we thought we were deserved it, we figured.)

On the way back to the convention center, we saw a huge billboard from Stockmen's hotel, with a giant picture of a cow on it and the words "Room for Fun."

"Bring your cow," I said, chortling.

"Plenty of room, bring your whole herd," Terrell added, holding her stomach.

"Suites available," I wheezed.

"Chapel open 24 hours," she said. We had to pull over.

Sure, it sounds juvenile NOW. Two grown, married mothers, sober and successful, on the job even, acting like a couple of 14-year-old girls. But it was a hell of a lot of fun. We stayed up late and talked about our husbands and our mothers and ate cake in the middle of the afternoon even when it wasn't anyone's birthday and didn't worry about getting anybody else's clothes ready or making sure anybody else was having a good time. When we met a guy from Voice of America, Terrell talked him into reading my columns while I pretended to be shy. When Terrell wanted an interview with Baxter Black, I kept an eye out and reported his whereabouts so she could stalk him successfully. And she never hawked into a dirty bandanna or mentioned Hillary Clinton or scratched anything in public.

I don't understand women who eschew the company of women so they can spend all their time working their wiles on men; don't like them much, even. They're giving their gender short shrift. And no matter how much you love a man, only another woman can dependably make you laugh until you're sick. And frankly, only another woman would want to see it happen.

Passin' time with Patsy at Sally Q's

By Steve Crump

I got me a nasty letter from a feller in Texas last week, complainin' about a column I wrote up a while back about cowboy bars.

"You spudkickers (except he used a different word than `spud') wouldn't know a real cowboy bar if the handle on the Coors spigot rared back and smacked ya'll between the eyes," he snorted.

Them's fightin' words, and we're not gonna step outside to do it, neither.

Honor's at stake here, and I'd stack up Idaho cowboy bars against any pile of empties west of the Pecos.

We got us some jukeboxes. We got vinyl. We got 2-for-1 longnecks on Tuesday nights.

We got Sally Q's.

Sally Q's is a roadhouse east of American Falls that purdy much sets the standard for cowboy bardom, you ask me.

To start off, it's got character, which means it's been in bidness since the end of World War II. So's a lot of the mud and the blood and the beer that have seeped right into the fabric of the place.

Second of all, it was country before country wore a $500 Stetson. Nobody at Sally Q's asks for a coaster, and if you're gonna slide into that booth in them Calvin Klein jeans, then I dang well better not hear you whine about the wad of gum stuck to your butt.

Third of all, Sally Q's rates about a 10 on the Patsy Cline Index of Heartbreak and Hallelujah.

Ya' see, before cowboy bars got all gussied up, the folks who drank there did so fer purdy much two reasons.

They felt real bad or they were figurin' on it.

The first group knew that cowboy bars provide company so's you don't have to cry in your beer alone. That's how come Patsy Cline got to be such a big star. Her voice always sounded like she was about half a beer away from givin' up on men altogether.

On the other hand, people who go to a cowboy bar feelin' good generally don't plan on stoppin' until somebody steps on their knuckles on their way home in the morning.

Patsy knew about that too. "Walkin' After Midnight," after all, ain't a song about insomnia.

It's about a place like Sally Q's, which don't put on airs. It ain't got no stuffed armadillos, padded bar stools or big-screen TV, and you damned well better be light on your feet if you're a-gonna order snooty bottled water or cappuccino.

It sets right across the highway from a phosphate fertilizer plant, which means it attracts a mix of blue collars and blue pickups. There have actually been brawls over whether it smells bad.

All of which has meant that drinkin' at Sally Q's has developed a kind of what you might call a protocol.

You walk in, shove three quarters into the jukebox, insult the barkeep, take a seat, eat handfuls of peanuts in the shells, pay for every third round, and leave before you get sick.

You don't run a tab. You don't talk religion. You don't leer at a waitress who's carrying a tray full of beer bottles, and you don't stare at the drunk at the next table unless you want him to rearrange your bridgework.

In short, Texas, you'll either have a good time or you'll answer to Sally Q herself out in the parking lot.

After Patsy gets through with ya'll, of course.

How to pack a gun like Uncle Darwin

By Steve Crump

I'm a-gonna git me a gun rack for the family station wagon.

OK, so it will look stupid back there with bags of potting soil and the fondue set my wife wants me to take to the Salvation Army.

But you can't be a *real* Idahoan without a gun rack.

You could look it up. Right there in the state Constitution after life, liberty and the pursuit of slow four-point bucks too dumb to hide in the underbrush is the provision that guarantees citizenship to anyone who packs his artillery in front of God and everybody.

Why? It's plain enough. Never met a man with a gun rack who wouldn't offer up his hand, look you straight in the eye and threaten to deviate your septum if you scratch his truck.

Us Idahoans believe in the eternal verities: a three-quarter ton pickup with four-wheel drive, a .30-06 and overdue payments on both.

A gun rack is install-it-yourself macho, serving notice on the world that you're not to be messed with. It implies that you're the kind of fella who'll shoot first and ask whether it was loaded later.

It's a powerful symbol of Western independence, tinged with the fear that the driver just might be a few longnecks shy of a six-pack.

Heck, if Betsy Ross had stitched a gun rack on our first flag, the Brits would have cleared out faster than government employees on a Friday afternoon.

Besides, you can tell a lot about the character of a man by what he packs on his gun rack.

The standard armament is a .22 with a scope, just the thing for pluggin roadside varmints or Iraqi paratroopers, but personally I think that's the wimp's way out.

I admire a man who packs guns for any occasion on a triple-decker.

See, you put your .22 on the bottom hooks for small mammals with an attitude, then above that you keep a loaded shotgun just in case Alfred Hitchcock was right about them birds.

On top goes the big gun with a clip, handy if a bull elk pulls up next to you at a red right or if a street gang selects your driveway as a location for a factory outlet.

Finally, you keep your .357 Magnum in the glove compartment in case you lose the key to the gun rack.

My uncle Darwin was a wheat farmer who carried a .45 automatic in his pickup just so a rockchuck wouldn't get the drop on him. Came over a rise in the road one morning and saw a dozen of the marmot varmints in the borrow pit, big as the Hole-in-the-Wall Gang waiting to rob the UP payroll train.

Darwin got so excited that he clean blew out his windshield.

With the insurance money, he bought himself one of them pickups with a rear windshield that opens up and he installed a gun rack right across it. The idea was that come duck season, he'd sit in the truck and stay warm, then crank open the window and blaze away first time a merganser came within range.

Uncle Darwin went out opening day and parked on a little hill overlooked the Bear River, but he didn't have long to wait. As he watched, the northern sky grew black with mallards closing in on him.

Real quietly and very carefully, Darwin opened the window and poked the muzzle of his old Steven's 12-guage into the icy November dawn. Then he drew a bead on a flight of about a half dozen nice fat ducks as they maneuvered for a landing on the water behind him.

And then ... well, did you ever try to aim a 14-pound shotgun through a three-tier tempered steel gun rack?

Darwin's insurance company replaced the rear window, but they wouldn't pay to take the buckshot out of the tailgate. So

Darwin took the tailgate off and had to carry his fishing gear up in the cab — on the gun rack.

Seems like most gun racks, sooner of later, end up carrying fishing poles.

Show me a man with a fly rod on his gun rack and I'll show you a man whose wife has locked his shotgun in the trunk of her Buick.

I wonder if I'll get it back when she sells the car.

Chapter 9
You can't sleep in on John Wayne's birthday

Stumped by the 4th of July

By Steve Crump

There's a stretch of Pegram Road just this side of the Wyoming border that used to be the most dangerous highway in the West.

It wasn't the traffic. It was the explosives.

Next to the road, you see, stood an old stump, and under the stump we hid our firecrackers.

Had to. Deputy Strubert was on the case.

The deputy patrolled a particular section of highway that led to Border, a Wyoming frontier hamlet that was a mecca for teen-age pyromaniacs in those days.

Wyoming, you see, took a laissez-faire attitude toward things that go boom in the night; Idaho barely tolerated sparklers. Back in 1966, that discrepancy meant that much of Border's economy depended on young Idahoans in battered pick-ups who'd cruise over in the weeks before the Fourth of July and stock up on fireworks.

We'd get up at four in the morning, drive the 90 miles from Pocatello and try to slip into Border and get out again before the deputy came on duty at eight.

Sitting in his black-and-white atop a bluff on the Idaho side of the border, Strubert could clearly spot his quarry. Most of the cherry bomb-runners were familiar to him anyway, so each time a vehicle with Idaho plates crossed the Wyoming border back into his county, he was ready.

"Been to Border, boys?" he'd ask, slowly surveying your rig. *"Y-Y-Yessir,"* you'd say. *"I mean, N-N-Nosir.* Been fishin`!"

"Fishin'?" he'd allow. "Catch anything?"

"No sir!" you'd exclaim, about the same time that Billy Foster, sitting in the seat next to you, would blurt out, "Chubs!"

"Chubs, huh?" Strubert would reply, scratching his chin. "I don't suppose you have 'em with you."

"No, sir!" you'd reply mock-sincerely. "We dumped 'em."

"I see. Then I don't suppose you boys would mind showin` me your Wyoming fishing licenses."

This was tricky since, technically, the deputy had authority to detain anyone he suspected of a fish and game violation.

"We didn't catch 'em in Wyoming, sir," Billy would volunteer after an agonizing silence. "We caught 'em in Taylor Creek!"

Taylor Creek was an unprepossessing trickle of meltwater that descended from Boundary Ridge, a 20-mile-long mountain that ran south along the Idaho-Wyoming line. It contained no known fish, a fact well-known to Deputy Strubert.

"Then you boys sure wouldn't mind showing me your *Idaho* fishing licenses."

Billy had his in his wallet. It had expired the previous December. The best I could do was a superannuated elk tag, from 1965.

"Well, you boys got two choices," the deputy would offer at last. "You can come along with me to jail or you can give me the firecrackers."

The fireworks were hidden in the rocker panel on the driver's side, or sometimes in the spare tire. Strubert knew this all along, of course, and the brightly colored packages were soon in his hands.

"If I catch you boys runnin' firecrackers through my county again," he said, "I'll call Billy's uncle."

Billy's uncle was the sheriff of Bannock County, where we lived, and a phone call from Strubert would effectively have grounded both of us through the second Nixon Administration. Worse, Billy's '56 Ford pickup would have quickly found itself consigned to Friendly Fred's Used-Car Sales.

Still, there's no force on earth quite as powerful as a teenager who has been told no, so late the following June, we found ourselves in Border again, but this time with a plan.

We arrived in the middle of the night, and at first light bought Wyoming fishing licenses, having first remembered to purchase their Idaho counterparts. We also bought $100 worth of fireworks and, before Strubert came on duty, stashed them beneath the stump beside the road.

Billy and I were in the process of transferring the firecrackers into a series of cigar boxes that we had taped to the underside of the cab when the deputy pulled up in his patrol car.

Still, we felt pretty safe. The cigar boxes were undetectable unless you got down on all fours and crawled under the truck, and Deputy Strubert weighed 275 pounds.

We answered all of his questions satisfactorily until he wanted to know what that odd odor was coming from beneath of Billy's truck as it sat there idling by the side of the road.

That's when a series of small explosions blew the floor out of the cab. Billy and I dived into the ditch just as the nearby stump blew up.

It seems some of the hidden cigar boxes had opened, spilling M-80s on the red-hot exhaust manifold and the tinder-dry grass underneath the truck. Soon the explosions found their way to our cache of fireworks.

Billy went to military school soon thereafter, and I didn't get to drive again until I was 22 years old.

Deputy Strubert, though, got elected sheriff. Guess he was an explosive stump speaker.

Erotic dreams

By Suzanne Huxhold

CLOVER CREEK- "Honey," I crooned into my husband's ear Valentine's morning. "I had the most exciting dream about you last night."

"Hmmm?" He lifted one sleepy eye in my direction. "Tell me."

"Well, I was going to take a shower, see..."

"Yeah?"

"...and I took off my robe and stepped into the tub..."

"Go on, go on." He snuggled closer.

"...and there was all new ceramic tile in the bathroom and you'd put in new grout around the soap dish..."

"Oh, geez, not this again."

"...and you'd installed a heat lamp and one of those towel warmer things..."

He was already out of bed, yanking on a pair of jeans. "I'm not listening," he chanted, "I'm not listening."

"Don't you want to hear the rest of my dream?" I asked innocently.

"You've had this recurring nightmare before, I think."

I flipped back the comforter invitingly. "I'll tell you about the new tub fixtures if you come back to bed."

He glared at me. "You know, my mother isn't the only person who occasionally wonders why I married you," he said, and stalked off to the kitchen to start the coffee.

Valentine's Day. Who gets it?

Valentine's Day comes at a very inconvenient time for Scott and me. He's right in the middle of calving season, feeding season, barn-repairing, fence-mending, bull-buying and tax season. And all I want is a little simple romance. Which would be well and good if he could satisfy me with flowers and the kind of underwear you'd never wear outside the house in case you got in a car accident and the paramedics thought you were some kind of pervert, like your mother warned you about, but he can't. Oh, he WISHES he could. One quick trip to town and it'd be all over with, his duties fulfilled for another year. Unfortunately for him, however, romance, for me, comes in the shiny-eyed, full-hearted form of home remodel. Flowers and underwear I don't care from dirt. But approach me on Valentine's Day with a socket wrench and a gift certificate from Gooding Lumber, and I'm yours forever.

Which is not to say I'm not a victim of spring, like any creature. When the Hallmark people voted in mid-February as Valentine's Day lo these many centuries past, well, their timing was perfect. But while in Spring other young women's fancy turns to matters of the heart, mine turns to re-facing the kitchen cabinets.

It's easily explained, of course. I'm a nester. From a long line of nesters. GEESE don't have the nesting instinct the women in my family have. Come Groundhog Day, we're ready to gather our twigs and horse hair and start something. After 11 years of marriage, my husband should understand this, don't you think? Has there BEEN a Valentine's Day since 1984 when I haven't bought paint? Have we EVER got through a February without the belt sander and the wallpaper paste? I don't THINK so. And yet, every year he tries to buy me off with lobster and jewelry. I get big flowery cards with his own words of love blah, blah, blah. I'll never understand men as long as I live. Don't they know what love MEANS anymore?

My mother called on the 15th. "What did you do for Valentine's Day," she asked.

I scraped at a paint fleck on the kitchen counter with my thumbnail. "Oh, nothing," I said moodily. "Scott made stuffed veal with poached pears and wore a tie and didn't eat lunch for the past three months so he could buy me some diamond earrings. Same old, same old."

"Oh, honey," my mother commiserated. "I'm sorry."

"Well, you know. Eleven years. The romance is gone, I guess. What did you and Dad do?"

"We ripped up the carpet in the den and refinished the floors."

Some women. They live in a dream world the rest of us can only imagine.

I smell a rat

By Suzanne Huxhold

CLOVER CREEK-So who isn't glad the holidays are over? Hands?

No one? Not one single one of you sorry to see the holidays over? Yeah, that's what I thought.

And I know why you're all happy to have all that behind you.

It's because your families are insane, isn't it? Isn't it?

Admit it. You think they're nuts. You think they're overly sensitive, crabby lunatics who buy cheap gifts and for some reason are still holding a grudge from that ONE Christmas when you ran over their dog.

I got news. They're all sitting in their living rooms saying the same stuff about you.

Everyone I talked to who had a nice season didn't spend any part of it with anyone who shared their gene pool. Everyone else swore they were going to Cabo San Lucas next year. Swore. Contacted travel agents on the 26th. They were not kidding.

My own family is not immune. And I consider us to be quite close. I turned to my husband on Christmas night, sated with prime rib and dear friends and good will toward men and whispered romantically, "I'm having such a wonderful Christmas," and he whispered back, "That's because we're not spending it with your family."

Sad but true. Sad but true.

And when I called my sister the next day, it became even truer. True as true can be. True like Gospel.

Because my cousin brought a rat to Christmas dinner. And call me crazy, but to my way of thinking, nothing spoils Christmas more than a big rat.

So my sister arrives at my grandfather's house on Christmas Eve. She's met at the door by my cousin. As my sister goes in for the Annual Christmas Hug, my cousin says, "Careful! I have Daisy in here."

Now, as Daisy is the name of my grandmother, who passed away last spring, and being that they were at my grandmother's house, my sister was disconcerted. Plus, she was wondering where "in here" was supposed to be. Just then, a rat head peeked out from the collar of my cousin's dress.

"And this wasn't any cute little white rat," my sister tells me. "This thing's half as long as my forearm. It looked like those big gray avocado rats we used to find in the trash cans when we were kids. I kept saying `Here, kitty kitty kitty', but she didn't think that was funny."

So Daisy, or "the rat" as I'll call it for clarity's sake, just so you don't get it mixed up with my grandmother, stays inside my cousin's dress for the afternoon, crawling around and doing God-knows-what in there, while everyone is pretending not to think this is extremely weird.

"Hors d'oeuvres?" my cousin offers genteelly, while the rat runs around in her hair. My sister looks down at the plates of prettily-arranged appetizers. The rat cage, where the rat lives when my cousin is naked, apparently, is sitting right next to the food.

"Uh. No," she whispers weakly. "Thank you anyway." I tell her what a rock she is. I might have thrown up right then and there.

Let me interject here to tell you a little about my grandmother. My grandmother was a very fastidious woman. She would have burned down her house before she let a rat in it. On Christmas Eve. In her granddaughter's dress. Named after her. Burned it right down, I tell you.

My cousin calls me about 10 minutes after I hang up from my sister. How was your Christmas? she asks. Fine, I say. How's your rat?

She doesn't think this is funny, either. She's sensitive about the rat, evidently. Daisy's fine, she replies, somewhat tersely.

Did SHE have a nice Christmas? I ask. And laugh like a loon.

Which was very rude, because, when all is said and done, I love my cousin. She is not merely my cousin. She's like a sister to me. If I had a sister who was a rat-carrying nut, that is. No, really, she's sweet and gentle and loves all living creatures.

Which are very nice traits - that I don't happen to share.

"You really have a rat named Daisy?" I ask. Because my sister's been known to pull my leg now and again.

"Yes," she says. "You see, I've really been missing Nanny lately and I've been thinking so much about her..."

"So you named your rat after her?"

"I LOVE my rat!" she protested. Which, of course, is true.

True as Gospel.

And you think your family has a screw loose? Reassess.

If you've never had a rat at your Christmas dinner, you're practically the Joseph's of Bethlehem in my book.

By the way, I'm making reservations for Cabo San Lucas even as we speak. No sense taking any chances for next year.

CHAPTER 10
IT GIT ME ABOUT 100 BUSHEL OF ROCKS TO THE ACRE

Love, honor and subsidy checks

By Steve Crump

Gotta go to eastern Idaho next weekend. Big wedding in the family.

Cousin Ernie, who's 41 and hasn't really found himself, is finally gettin' hitched to Bernadele Bowersox, who is not only a successful businesswoman but thinking about running for the local weed board.

Now in any other farm family, such a mismatch might seem an omen of a combine wreck of epic proportions, but not in mine. Ernie, you see, is Uncle Einar's youngest boy.

Einar was a first-generation Norwegian immigrant who worked 160 acres of hard red dust in the Bear River Valley.

When the New Deal finally percolated down to rural Idaho, the county agent went to see Einar and explained to him that the government would pay him more for not growing wheat than he could earn for planting that spring.

That thought went down pretty hard with Einar, a proud man with scruples. But the Massey-Harris tractor needed fixing, Aunt Floradora wanted a roof on the house, and there was second mortgage on the chicken coop.

So Uncle Einar signed the paper and, in due course, received his first set-aside check in the mail.

Floradora found that same check, years later, after Einar had died. He could never bring himself to cash it.

She said it was hidden among gilt-edged stock certificates and treasury bonds in their safety deposit box. For Einar, you see, had prospered, despite the fact that he stuck to principle.

That's because Einar took his second set-aside check and went to Reno, where an inside straight was waiting.

Now Einar's kids own 20,000 acres of Idaho and a half block of Hawaii. The eldest son of the county agent is one of their many accountants; they pay him to just to deposit their CRP checks.

And Bernadele is that accountant's daughter.

Is this a great country, or what?

Just a big ol' girl

By Suzanne Huxhold

CLOVER CREEK — When it comes right down to it, I'm just a big old girl.

I try to be tough. Macho. I try to do the guy thing. But I just don't have the testosterone.

We were moving cows last week across the prairie. Not our cows. Our cows don't deserve to live on the prairie. Our cows have to live on the godforsaken desert, and count themselves lucky.

No, these were prairie cows we were moving. Prairie cows are different than desert cows. First off, they won't eat sagebrush, chaparral or rocks. They like grass, oddly enough. Green grass.

Desert cows don't believe in green grass. They're atheists that way. If they've never seen it, they don't believe it exists.

And prairie cows like water. Big pools of it. They like their water just lying around, too, close to where they live. If you can imagine. They won't hardly dig at all.

And prairie cows like to move. Fast. When you open a gate for a prairie cow, she'll just head right out. At a run.

I didn't know this previously. I thought prairie cows moved like desert cows. Desert cows KNOW there's nothing better out there, so they're never in any hurry to get anywhere. What's the difference between one lava rock, cheat grass hellhole and another?

Nothing. And desert cows know it.

But prairie cows, apparently, are moved OUT OF lush grassy fields into other lush, grassy fields at regular intervals. It's a whole system they got up there. It's called good feed. I personally don't understand it, but it seems to work for some people.

Anyway, we're moving these prairie cows. And the foreman asks me to ride point.

Now, I practically laugh at this. Point? Why not just set me under a shady tree and give me a book?

Point is the easiest place to be. You eat zero dust, you don't have to stare at cow's butts all day, and you're the first one in for lunch. I love point.

Not that I ever got to ride point before. I'm always at the back. With the kids and the calves. I always eat the most dust, stare at the most butts, and eat cold food at the end of the day.

So I happily take my assignment. I walk my horse out into the road with a smug smile. Point. I want to hug the foreman.

Then I look back. And am somewhat alarmed by the speed at which the cows are approaching. They're still coming through the gate a half-mile down the road, but they seem to be getting awful close, awful fast.

I nudge my pony to a trot.

The cows take this as some sort of sign. Now they're barreling toward me.

"Uh, whoa," I say, half to myself.

It's like a siren's call, this small sound. They lunge for me.

"Hey, stop!" I yell, putting my hand out like a school crossing guard.

They delight in this and redouble their efforts to get past me.

By now I'm criss-crossing the road at a canter, trying to keep them behind me. And I'm yelling.

"Help! Help!"

Just like a big old girl. It's humiliating.

More humiliating is the fact that the foreman, seeing my distress and obvious incompetence, has sent his MOTHER down the road in the pickup to get ahead of the cows. His mother! Can you imagine my shame? She flies past me in a cloud of dust just as the whole blooming herd breaks around me on either side and dashes for freedom. Or better grass. I can't tell.

I'm in disgrace. One cowboy gallops past me on my right, another on my left. They turn the cows just as they're about to hit the road for Mountain Home, never to be seen again. I've cost everyone an hour, I'm told, and I'm banished to the back with the rest of the dust-eating, butt-staring, late-lunch-getting crowd.

Who knew prairie cows could run 30 miles an hour? Desert cows can't move 30 miles in a week, unless you try and fence them in.

Then they can practically levitate.

I'll know better next time, I told the foreman. He just looked at me.

"So will I," he said.

Good help

By Suzanne Huxhold

CLOVER CREEK — "Chuck and I need you to help pick up pipe this afternoon," my husband said to me yesterday.

"Me?"

"Yeah, you."

"Pick up pipe."

"Yes."

"You want me to pick up pipe."

He glared at me, vexed. "Are you not understanding English today?"

"I am. It's just that you never ask me to help you guys. I always have to beg to be included."

He couldn't meet my eyes. "That's not true. I have to go now."

"It is true," I insisted. "You call Bill or Mike or somebody, but you never ask me."

He sidled out the door. "I don't know what you're talking about. Be at the tractor after lunch."

It is true, you know. They never do ask me to help. Oh, sure, I'm usually included when it's time to call the ranch owner and tell him we need money to buy a crank for the swather, but other than that, I get ditched. I really don't understand it.

I was at the tractor right after lunch.

"Ready?" Scott asked.

"Ready. I just have to run Joseph's bike over in the van. I'll be back in 10 minutes."

They waited for me. Wasn't that sweet?

"Ready now?"

"Ready! Oh, wait, I forgot sunscreen. Be right back."

I put my sunscreen on, and got a hat. Then I drove the tractor out to the hay field. It seemed slow, so I revved it up a little.

The little rabbit on the right is for fast, I remembered. I bounced Chuck right off the pipe wagon, though, so Scott reached in and jammed the throttle back toward the little turtle.

"Oops! Sorry, Chuck," I said. He waved at me in that cute way of his, with the one finger pointing up. Chuck just loves me.

When we got out to the fields I ran over a section of pipe, but luckily it was the plastic kind so I'm sure it won't be too expensive to replace.

Then while I was driving along I couldn't get the speed exactly right so the darn tractor kept stalling, but everyone thought that was funny so I didn't feel too bad about it. Scott kept shaking his head. I know he must have been just laughing and laughing.

"Are there always this many mosquitoes?" I said at one point, and not in a whiny voice at all. But the tractor was pretty loud and they must not have heard me.

I got kind of bored after a while so I thought it might be fun to pick up pipe. But did you know those pipes usually have WATER in them? And they're HEAVY. So I quit after the first one.

"I love farming," I told my husband a little while later. He said something that sounded like "I could kill you right now," but I'm sure that was my imagination.

I had an allergy attack right after that, so we had to go all the way back to the house so I could take a pill and get some tissues and put a meatloaf on. Scott came in and called Bill. I don't know why.

Everything went pretty smoothly after that, except I drove the tractor into the irrigation ditch. But Scott didn't say a word about it. Not one word. He just helped me to the ground, climbed in and backed it out, grinning the whole time. He must have thought it was pretty amusing, because I never saw him grin like that before.

His lips were curled way back and I could see all his teeth.

When I asked if I could help them get pipe off the north fields today, they said yes.

But coincidentally, Bill came out this morning so I didn't get to help. But now that Scott and Chuck have seen first hand how helpful I can be around here, I'm sure they won't be ditching ME any more.

Comemos Sus Patatas!

By Suzanne Huxlold

CLOVER CREEK — We went to Central America a couple weeks ago. Don't ask me why.

It was a whim. My mother said last year, "Why don't we go to Central America?" And since we couldn't think of any reasonable answer to that, we went.

I planned the trip. I won't get to do that anymore.

Because believe it or not, I think they hold me responsible for the 36-hour plane trip and the three hours in customs and the body cavity search. (Apparently, drug trafficking is a problem in Central America! Who knew!?)

I believe they blame me for the tropical storm, the hurricane winds, the driving rain. The sand fleas, the reef-grass lice, the poison coral. The four adult sharks that lived in the swimming lagoon outside our bungalow.

I believe they blame me for the fact that our introductory presentation at the resort included warnings about coconuts falling from the trees and killing us, about how if we were late for meals it was our tough luck, about how too many hot showers would add to our bill (which wasn't a big problem, as the hot water tank in our room held about three cups of hot water at a time. We started heating water in the coffee maker by the third day). And about how we had to be careful not to get sick or hurt because we were irrevocably stuck on the godforsaken atoll for seven more days. They actually used the word godforsaken, which was not anywhere in the brochure.

Our hosts were French, by the way, if that gives you any indication as to how naturally friendly they were. We were 15 minutes late to breakfast one morning and they nearly had us all flogged. I asked for extra rolls at dinner; you'd have thought I was Oliver Twist. "More rolls!? MORE!?" My request for shampoo nearly got me thrown into the open sea.

I think my family believes I had something to do with the 30-foot seas. And the three-hour, 72 mile, near-death experience of crossing those seas. At night. Each way. With 20 Central Americans who spent the trip alternating between barfing out the back and screaming prayers in Spanish.

Oh, my family never actually said it was the worst vacation of their lives, but I got the gist when we had that unfortunate jet engine failure on the runway at the Miami airport and were stuck inside the plane with no air, no bathrooms and no movie for two hours while the guy crammed in the seat next to us, who'd cut three of his fingers off in a skill saw that morning moaned in his sleep, and my mother stood up in the aisle and started telling the flight attendants to shove me out of the plane because I was obviously the travel equivalent of the black angel of death!

Well, you don't have to knock me over the head! Next year they can just get a travel agent.

But something interesting did come of my Stephen King vacation nightmare. Besides the malaria shots.

Idaho potatoes are everywhere!

It is a phenomenon of the universe that even in the jungles of Central America, where Mayan ruins rise from ancient earth and people still carry stuff around in baskets on their heads, you can tell a Spanish-speaking villager you are from Idaho and they will say, "Carumba! comemos sus patatas!" Which either means, "Hey! we eat your potatoes!" or "Cripes! where did you get all those sand flea bites on your neck?"

We ate Idaho spuds every day. Little Hagerman red potatoes in parsley butter, Burley bakers, Gooding home fries with Wendell onions. We felt entirely at home. Even barracuda meat in lime and mango sauce tastes surprisingly homey when you get a big old blob of mashed Idaho potatoes with it.

We were well-received by way of this potato connection, incidentally. The other people on the godforsaken atoll with us were from New Mexico. We laughed at them behind their backs. What the hell has New Mexico got on their license plates? The sun? Like the rest of us don't ALSO have the sun?

I tell you, those Central Americans can relate to a good potato farmer. One of the guys on the island grew up on a cattle ranch, too, so we were practically natives. Talk a little bull, eat a few spuds; we were like family with these people.

We sent them all Idaho Potato Growers hats when we got back, and those little candy Spud Bars.

We'll have to go back someday to see if they liked them. I'm sure my family can't wait.

CHAPTER 11
PICKUPS AND OTHER LAWN ORNAMENTS

So you want a driver's license, eh?

By Steve Crump

My oldest son will be 15 in October, so it's about time to find a flat country road with not a lot of mailboxes.

Oh, sure, he'll take driver's training next year, but there are some things that you just can't learn from a superbly qualified, incredibly patient instructor with high blood pressure and a little facial tic.

He may teach you how to parallel-park a Ford Festiva between two sets of cones, but real drivers have to know how to handle a '61 Chevy pickup.

I'm not sure why. They just do.

Maybe it's to experience firsthand what a stick shift sounds like when it's grinding gears into a fine metallic powder, or to learn to navigate through a cracked windshield in the driving rain when the wipers are broken.

Perhaps it's to feel the wind pouring through the floorboards or the rush of adrenaline when you slam on the brakes and put your foot through the firewall.

Or maybe it's because of the Curtis Jenny.

Pilots used to learn to fly on the Jenny, which was made of stick, string and fabric, and not a lot of each.

It hummed along at treetop level at about 75 mph, just an engine cough away from becoming a weather vane on somebody's barn.

Wannabe aviators got a real sense that Mother Nature wasn't all that thrilled about their being there. They also learned all three verses of "Nearer My God to Thee."

That's what a '61 pickup did for the young driver: gave him a real sense of his own mortality.

I took my first spin in my Uncle Farron's Chevy. He was an unflappable sort of fellow, a dry farmer from Soda Springs who chewed snoose and spent his winters playing roulette in Reno, but the gambler in him made Farron the perfect driving instructor.

He'd sit in the passenger seat, pull a tin of Skoal from his shirt pocket, stick a pinch between his lip and his toothless gums and then spit out the window. "Don't hit the chicken coop," he'd say simply.

That was no mean feat. A wimpy sixth-grader steering an old truck is not a pretty sight.

One morning, Farron and I headed into town for groceries, and as was his wont, he let me drive.

Went into the borrow pit and up into the winter wheat a couple of times, but I didn't get into real trouble until we came to the sharp curve front of the Widow Duncan's.

Mrs. Duncan supplemented her Social Security by selling eggs, and she had an ice house out by the well and next to the chicken coop.

But the odd thing about Mrs. Duncan's place was her clothesline. It stood close to the road and squarely astride the lane to her house, so that when you went to visit her, you had to drive around her laundry.

A couple of times in the previous year, Uncle Farron had cut the curve too close and sprayed mud and gravel on Mrs. Duncan's delicates. She had a word with Aunt Edith about that, so Uncle Farron was under strict orders to mind his daisies and dosies.

"Slow down here, Stevie," he said that day. "We don't want to cut it too cl ..."

It was already too late. I'd made a great roundhouse turn of the steering wheel, which of course sent the truck careering toward the other side of the road and a large patch of lava rock.

"Turn back!" Uncle Farron shouted.

I turned back, all right, overcorrected, left the road and ran squarely through Mrs. Duncan's clothesline, taking three sheets and a comforter with me.

"Don't hit the chicken coop!" Farron yelled.

You'll be proud to hear that I didn't. Glanced off the well instead and took out the ice house.

Didn't hurt the truck, but can't say the same for Uncle Farron after Mrs. Duncan caught up to him with a tire iron.

I don't know that they ever got the egg shells out of the radiator, but the whole business taught me a rather valuable lesson about driving that my own son will never get in a classroom.

It's that the wise instructor lets his student make his own mistakes — from a safe distance.

Borrow your pickup next weekend?

The Jerk

By Suzanne Huxhold

CLOVER CREEK — Did you ever see the movie "The Jerk?" Steve Martin plays this poor, dumb guy who makes a lot of money then loses everything.

As he's leaving his house, destitute once again, he grabs a lamp off the table and says, "I don't need stuff. All I need to make me happy is myself. And this lamp." Then he grabs a phone book or something (I saw this movie a long time ago) and says, "All I need to make me happy is myself and this lamp. That's all I need. Plus this phone book."

He goes through the house, grabbing stuff, telling himself all he needs to make him happy is more stuff. "All I need to make me happy is myself and this lamp. That's all I need. Plus this phone book. And a shoe. This one lovely shoe. That's all I need. If I have this lamp, this phone book, and this shoe, I'll be happy.

Plus, I need this oven mitt. Myself, the lamp, the phone book, the shoe and the oven mitt. And that is all I need! And maybe this plate."

I relate to this movie. I LIVE this movie. With my husband as the jerk.

Actually, the ranch is the jerk, Scott's just providing the stuff. All he needs for the ranch to make money is more stuff.

"We need to buy a motorcycle," he says to me last night. He's reading the classified ads again. I don't even bother to look over. I know the routine.

"We do not."

"No, we do! Think how much money we'd save."

"By buying a motorcycle?"

"Yes! I could ride it out to change water and fix fence. Think of the gas money. The wear and tear on the pickup!"

"How much?"

"How much which?"

"How much would we save?"

"Oh, five, six hundred bucks, overall. At least."

"And how much is the motorcycle?"

"Um. Used? About five, six hundred bucks."

"Huh," I say. I believe that's enough of a statement. But he's fixated.

"I've been thinking about this motorcycle thing for a couple years now. I really think it's a sound business idea."

"You've been thinking about it since our neighbor Tom got one last week."

"And look how much money he's saving!"

I just grunt at that.

"Look," he presses his point, "all we need is just this motorcycle. This motorcycle is all we need. The ranch will go into the black if we just get this motorcycle."

The desperate look in his eyes is really heartbreaking. I can see how happy it'll make him if I give in. But really, a precedent like that? You never know where it will end.

Pretty soon he'll be walking through the tractor dealer, dazed, saying, "All we need to make money is this one 1996 Case backhoe. That's all we need. This 1996 Case backhoe and this new John Deere with air conditioning and a stereo/CD player. And that's all. Plus this $50,000 swather. If we had this 1996 Case backhoe and this John Deere with the cool interior and this $50,000 swather, we'd have all we need for the ranch to make money."

I shake my head. "You know we can't spare $600 for a motorcycle right now. We've got a summer rider coming on, property taxes are due in June and we just paid the insurance."

He wheedles. Cajoles. Sulks. But I hold strong. And keep the ranch checkbook under my pillow. Pretty soon I hear the newspaper rustle. He's turned the page and I breathe a sigh of relief. Until he says, "You know, they've got a bass boat in here for $3,000. We really need a new bass boat."

"How in the world will a new bass boat help the ranch make more money?" I ask.

"Oh, it won't. But a new bass boat is all I need to make ME happy. All I need is just you, Joseph and a new bass boat. Plus, maybe a better car. You, Joseph, a new bass boat and a better car. With air conditioning and a stereo/cd player. That's all I need.

And maybe a snowmobile."

I'm sleeping with the personal checkbook under my pillow now, too. It's lumpy, but better than bankruptcy.

Twin Falls drivers don't know the meaning of speed

By Steve Crump

Twin Falls has the slowest drivers in the world.

And we're not talking just the high-tech post-industrial Western world, either. There are motorists in Sarajevo clanking along in two-stroke Yugos — vehicles that can't go from 0 to 60 before the warranty runs out — who would wax the average Twin Falls driver in a drag race down Second Avenue.

And it's not just the older drivers in Twin Falls who are slow.

If you doubt that, check out the drive time down North Blue some Saturday night in mid-August.

Oh, sure, we have our speeders, but they're from Utah.

The true Twin Falls motorist owns a vehicle that hasn't been out of third gear since it left the Hyundai test track in Pusan.

While Twin Falls' city street pattern won't soon be copied in Southern California, let's be frank: It's doesn't take an advanced degree in quantum mechanics to drive to the mall here.

But try approaching the speed limit on Filer West or Eastland any morning. We're the only town in Christendom that can create a traffic jam with two cars.

My friends, we live in a city where the right turn is an art form, executed in slow motion over about a mile and a quarter and a day and a half.

Responsible people out there — family men, God-fearing women — are driving 22 mph. There are drivers under 30 who've never been over 30.

In the interest of our brake linings and our blood pressure, let's make it a community project to stop driving like we're from Twin Falls. Here are some helpful hints for bringing the town up to speed:

- Ladies, don't buy him that gun rack for his pickup this Christmas. Guys with gun racks in their pickups drive slowly to show them off.
- When Mom says she's just going to the fabric store, hide the car keys; she's going to cruise Addison East without once taking her foot off the clutch.
- Do not feed, diaper, discipline or attempt to assault a child while the vehicle is in motion.
- Remember, it is permissible to drive a motor home from Twin Falls to Magic Reservoir faster than 8 mph.
- It is not necessary to come to a complete stop before making a right turn into the Fred Meyer parking lot.
- Teen-agers, when cruising North Blue do not park in the northbound or southbound lanes.
- Do not discuss politics, religion, money, sharing household chores or Jerry Springer while driving.
- Remember that any vehicle displaying a bumper sticker reading, "This car brakes for all garage sales" means exactly what it says.

CHAPTER 12
SEED CAPS AND THE ESSENCE OF HAUTE COUTURE

Plumbing nightmares

By Suzanne Huxhold

CLOVER CREEK — Remember the column a while back about swearing? About how cows make a person crazy? About how your sainted grandmother would turn into a cussing, spitting longshoreman with Tourette's Syndrome if she had to work cows for an afternoon? Remember that?

O.K., add plumbing.

Plumbing would also turn your sainted grandmother into a cussing, spitting longshoreman with Tourette's Syndrome. Plus, she'd stink like a swamp.

We've been living a plumbing nightmare for years, a nightmare we didn't even know about until this week, which as you can imagine makes it worse. Much, much worse.

"You know," I said once the weather warmed up a little this spring, "it always smells so musty in this house."

"`Musty' how?" my husband asked.

"You know, moldy. Wet, swampy, stagnant, putrefied "

"Yeah, O.K., I get it."

"Why do you think that is?"

"Dunno."

But to help, he checked the traps for decaying mice, helped me dig snack food out from under the couch cushions and put his cigars in the freezer. Didn't work.

Then, last week the sink in the kitchen backed up.

"Honey, the sink is backed up."

"'Backed up' how?"

"You know, clogged. Blocked, congested, obstructed..."

"Yeah, O.K., I get it."

"Why do you think that is?"

"Dunno."

But to help, he snaked and Dranoed and flushed and cleaned.

Didn't work.

The water, chocolate-colored with a layer of scum on top, sort of like a plumbing latte, stayed in the sink. Very appetizing. We ate out while he pondered the problem.

"I'll have to go under the house for this one," he told me the next morning. Being innocent, I said, hey, sounds like a good idea. You do that.

We should have just moved. I see that now. Him yelling "Oh my God! Oh my God!" from underneath the house was my first clue.

Apparently, from what I could gather between the swear words, some of our pipes aren't connected to anything. Like the septic tank. Or the outdoors at all for that matter. They may have been at one time, but that whole idea of our stuff going from our drains to someplace else just doesn't apply any longer.

Our stuff - and again, I just got bits and pieces of this, muffled by floor insulation and punctuated by cussing as it was - evidently drains right from our pipes into our crawl space. Both bathroom sinks drain there. So does most of the water from the shower.

"Ew, no wonder it smells musty," I said when he screamed this news up to me from the hole in the closet.

"Yeah, that's our biggest problem, Suzi," he said rather nastily. "The MUSTY smell."

There was other bad news, as well. The mud was bad, though most of it is on my carpet now so at least the crawl space is fairly clean. And there was some exposed wiring. Scott says the same **&$%$#@ guy who did the *&*%$#$% plumbing did the *&^%^%$# wiring! Which I gather is bad. And did I mention Scott is 6 feet 2 and the crawl space is 3 feet tall? Scott mentioned it. More times than you can imagine.

I didn't make it any easier on him, frankly. On one trip up from the "bowels of hell," as he began to refer to the crawl space, I said,

"Um, Honey, I hate like heck to bring this up, but the refrigerator doesn't seem to be running. Did you hit some wiring or something while you were down there?"

He got down on his knees and jiggled the plug. The damn thing started right up again.

"OOh, thanks!" I said brightly. He just glared at me.

And once when the phone rang I yelled down the hole, "Hey, can't you get the phone? I'm trying to watch TV up here!" He told me later he didn't think THAT was funny at all.

It took him most of two days and two trips to town, but he got everything re-plumbed. It'll take me two days and two trips to town to get the house clean again, but at least we won't have the musty smell.

Of course, if we do ever have that musty smell again, I'm planning to ignore it. That, or move.

It's time to go Idaho bonnet shopping

By Steve Crump

Lotta tough decisions get made every spring in the Magic Valley.

Do you plant bean seed or grain? Are you gonna get a better price for spuds or beets? And what kind of baseball cap you gonna get this year?

You city slickers scoff, but only because you buy your caps at places with 9-foot cardboard cutouts of Charles Barkley standing next to the door. No *true* Idahoan pays money for a hat.

Caps, you see, come in two categories: the kind you get for free and the kind you get free for test-driving a lawn tractor.

The first kind you get by showing up at your neighbor's, who works for a seed company, along about the end of February wearing a cap bearing the logo of a rival seed company.

If you don't happen to have a cap from the competition, just wear your old John Deere hat that you had on last summer when you fell into the combine.

Also helps to sit down at your neighbor's kitchen table, take off your old hat and scratch your thinning pate a few times. Ain't subtle, but it works.

"Say, Harlow, looks like you could use a new hat this year."

"Me? Shoot, only had this thing since my senior year."

"Well, go on and take one of my new ones. You can always save it for church on Sunday."

"Well, Bill, I'll go ahead and take it and give to Martha. We were ice fishin' at Magic last winter, and her hat fell in the hole."

Another sure-fire way to get a new hat is to walk into a farm implement dealer and pretend you've just about closed a deal with the guy down the road to buy a new seed drill.

"Tell you the truth, Fred, I just don't think anybody's gonna beat Bob's price. I'm just here 'cause you give me such a good deal last time."

"What's your hurry, Harlow? Set down, have a cup of coffee and we'll talk about it. Say, looks like you could use a new hat."

"Me? Shoot, only had this thing since my junior year."

"Well, go on, take one of my new ones. You can wear it to Jackpot on Saturday night."

"Well, Fred, I'll go ahead and take it and give to my boy. His mother give him a real bad haircut last weekend."

Of course, the big disadvantage of gimme caps is that the selection isn't great, so sometimes you have to negotiate.

"That's a fine lookin' pickup, John, but with bean prices where they are, I just can't swing it now."

"Did I forget to mention your $500 factory rebate? I'll even throw in a new hat."

"Well, I've only had this one since my sophomore year. What color?"

"We're giving away these real nice red 18 Wheels and a Dozen Roses hats right now."

"Say, that's about the color my neighbor Jeff's new pickup. He got a real good deal over at that place over on . ... "

"Say, Harlow, why don't you take my new hat? I sure don't need it settin` here at my desk figuring up your low payments."

If none of that works, go ahead and test-drive a lawn tractor, but have the dealer bring it over to your place.

"Ted, I'm real interested in that tractor, but I gotta tell you I want to see it run first. Bring it over Saturday morning. I'll be trimmin' the hedge. In the sun. Outside.

"What's that? A new hat? Well I got kinda used to havin' a sunburned scalp and I hate to give up a hat I've had since my freshman year. But shoot, bring it along. I'd sure hate to get sunstroke sittin' in the hammock watchin' you mow my lawn."

Nice feed sack; Is it new?

By Suzanne Huxhold

CLOVER CREEK — Vanity is a funny thing in a farm wife. You'd think, with all the disgusting stuff we do day after day - and my publisher has threatened to fire me if I name one more disgusting thing we do day after day, so I won't - we'd be beyond it.

I read an interview with a movie star not long ago in the checkout line in the supermarket. She was going on about how she's only had very minor plastic surgery because, after all, she doesn't want to get caught up in all that false value stuff in Hollywood.

One implant job (which implies, I'm assuming, TWO actual implants, unless there were some serious defects in the first place), an eye tuck (which is way more disgusting than anything that happens on this farm, I'm telling you right now) and some liposuction, AND THAT IS ALL! That's what she said. AND THAT IS ALL!

Well, good thing you aren't caught up in any false value stuff, baby. I mean, we can ALL understand the silicone and the liposucking and cutting off all the flesh around your eyes. God forbid you should have to go through life without breasts that make you tip forward, stick legs, and eyes that won't blink. (Actually, now that I think of it, my chickens all look like that. I should get them an agent.)

Farm wives just don't have the time for that kind of vanity.

Oh, I know one dairy wife who wears false eyelashes and big hair out to dinner, but she's definitely the exception. And I suspect she has some sort of pact with the devil, frankly, to look that good after she's spread manure all day and bottle fed 150 calves, but, you know, who's to say? Pacts with the devil are pretty hard to prove.

I've lived quite happily without vanity for 10 years now. I work hard. If I have time for a shower and lipstick before I go for a cheeseburger at the Ox Bow, well, I count myself lucky. I brush my teeth twice a day, wear sunscreen when I remember to, and shave my legs when my husband asks if I'm ever going to again, and that's about all I have time for. I've been perfectly happy with that setup. If I wanted to worry about my looks all the time, I wouldn't have a job that requires me to carry a shovel.

Recently, though, I was faced with the fact that, although I work like a man, I still have the brain of a female. A vain, silly twit of a female. A movie star kind of female. But with my own breasts, of course.

I was helping out in my son's first-grade classroom a couple weeks ago. To win their confidence, I got down on my knees, eye level, and gave them all my most brilliant smile.

"Eeew," one of the little darlings yelped, "you need to brush your teeth."

My smile dimmed slightly.

"Mrs. Huxhold, you have a yellow tooth," another shouted. I clamped my mouth shut.

"Joseph's mom has a big yellow tooth, Ms. Morrison. That means she has a cavity, doesn't it?"

This went on for several minutes, each fascinated little ratchild coming over to view my "big yellow tooth" like it was part of some dental science experiment gone hideously wrong. They pried open my mouth with their sweaty, grubby little fingers, viewed the offending tooth with their beady, squinting little eyes, and gave me lispy lectures about dental hygiene.

I hate children. At least I HAVE front teeth, I wanted to tell them.

In desperation, I explained how I broke the tooth when I was a child, and the bondo stuff the dentist had used back in 1970 had just discolored a little, that was all. I have excellent dental hygiene, I cried. Not a single cavity in my entire head, I shrieked. Look. Look. They didn't care. Children are so cruel.

I succumbed, I tell you. I folded like a pup-tent in a windstorm. Vanity came screaming back to me like a long-lost sister, slapped me upside the head and yelled in my ear, "You can't stay on that farm forever, sweetheart."

I went to the dentist, let him cram all his 20 or 30 fingers in my mouth for two hours, saw smoke come out of my own mouth, and came home with a perfectly white front tooth. I looked FABULOUS! Damn little first graders, I thought. Wait till they see THIS!

I went to school the next day. Smiled like a mental patient at everybody who crossed my path. Not a single child, nor man nor woman, noticed the $150 worth of new bondo on my tooth. So much for vanity. It's a waste of time for farm wives. I'm sure of it now.

Of course, if I got some nice implants.

Don't change tradition; change your chaps

By Steve Crump

I didn't see much of my half brother Pete while I was growing up.

He was 23 years older than me, and our dad died when I was young.

But it seemed to me that every time Pete showed up at a family reunion, he got a little shorter.

Not that the man was tall to begin with – 4 foot, 18 inches, mebbe – in a new pair of Tony Lamas.

But his stature wasn't helped by the fact that he rodeo'd, or worse, that he was a bullrider.

Purdy good one, too. Placed in the top 10 twice at the old Madison Square Garden national rodeo championship and four times at the National Finals Rodeo.

He rodeo'd from 1945, when he was 16, until 1966. That's – what? – mebbe 10,000 tumbles from the back of a half ton of whirling testosterone. Not counting the times he got stepped on.

Every time Pete got his leg broke – and that's the correct verb form among cowboys – his bones would knit a little tighter and Pete would get a little more compact, until he achieved The Walk.

You don't see much of The Walk anymore because most professional cowboys get out of rodeo before they get that busted up. The money's good enough to do it nowadays.

But when you make a career of gettin' kicked by sides of beef, you eventually start to plant your feet in your boots a little pigeon-toed every morning.

And you walk around with the moving parts in your knees grindin' together like a Ford engine with a hole in the oil pan.

Must hurt like hell. Hurts worse to watch. But The Walk is a badge of honor that any decent citizen who's ever raised either too many callouses or too much hell can respect.

That's why I'm pretty sure that Pete, who passed away a couple of years ago, would have somethin' to say about the pink chaps.

A bronc rider wore them at the Magic Valley Stampede – aka, the Twin Falls County Fair and Rodeo – last week. Even the rodeo clown made fun of him.

The clown, heir to an honorable, silent profession, could do that because he was a'wearin' a Madonna-style headset and a'swappin' lies with the announcer over the PA.

They played a Virginia reel during the bareback rides and replayed every second of it on a Diamondvision TV screen.

Pete's game just ain't the same.

I 'spose that's progress, but I don't know. Rodeo used to be somethin' you could count on, unchangeable as a $10 bill in a saddle tramp's jeans.

Folks used to say that if you'd seen one rodeo, you'd seen them all, and fellas like Pete would take that as a compliment.

"Rodeo's no different," he'd explain if he were here today. "It's a crime to try to gussy it up.

"Look, them Spaniards don't go to a bullfight to watch the bull chase the matador out into the parkin' lot. They go because the matador is gonna win – one way or another – 10 times out of 10. Ain't never been a toro that didn't end up as a beefsteak."

He'd pause to stub out his Camel straight and continue.

"Same deal with rodeoin'. Folks come to watch some damn fool fall off. It might happen sooner or it might happen later, but it's still gonna happen. It's what's right. It's tradition.

"Gimme a ring announcer who's gonna tell the same stories, a timer who's always a full second off the switch, and a rodeo clown who'll squat down inside his barrel, take a charge and wind up in the next county.

"And when he gets up, I wanna hear the announcer say, 'Give 'im a hand, folks, 'cause that's all the pay he's gonna git tonight.'"

Then Pete would stand up, hitch up his Wranglers, and launch a stream of tobacco juice onto the lizard-skin boots of the drugstore cowboy a'settin' at the bar.

Oh, he don't think much of your pink chaps, neither.

What the hell is that?

By Suzanne Huxlold

CLOVER CREEK — I had a giant pimple last week. I know that's a pretty personal thing to tell you, but it's nice we can share my misery, don't you think?

Besides, if you had been in Gooding County from last Monday to last Thursday you would have seen it anyway, so what's the difference?

It was a doozy. Right on my chin. People talked to it. I didn't make eye contact with one person for four days. I globbed makeup on it whenever I left the house, but that just made it look worse. Imagine someone trying to spackle over Kilauea.

So I walked around looking thoughtful. "Hmm," I said at every opportunity, tapping my fingers over the horrible disfigurement. "Now let me see." Which works O.K. for about five minutes, after which time everyone thinks the pimple on your face is the LEAST of your social problems.

I called a friend of mine in Boise, lamenting. You know what she told me? She told me when she gets a pimple, she just puts black eyeliner over it and pretends it's a mole. Like Cindy Crawford.

"Dab a little eyeliner over it and go out," she advised. "No one will notice."

Nope, I said. It'd never work.

Sure it would, she said. Why wouldn't it?

Well, aside from the fact I don't think black spots on my face are all that much more attractive than red ones, I could never pull it off. I live in a small town. A very small town. New facial moles would not go unnoticed.

"Cripes, Suzi!" I can imagine my neighbors saying, "Did that thing grow up overnight!?"

"You better get that checked, honey. I had a heifer that got one of them big old black things on her face like that. Died in three days."

"Honey, between that and the zit you had last week, you need to get yourself a dermatologist."

But what they'd say to my moley face would be nothing to what they'd speculate about when my back was turned.

"Did you see that thing on Suzi's face?! What the hell IS that?"

"I don't know, but I reckon she's got something catching, if you know what I mean."

"Didn't the bad witch in the "Wizard of Oz" have a mole in exactly the same spot on her chin? And didn't I hear there was a suspicious animal sacrifice out there on Clover Crick not two months back?"

A person can't go up a dress size in this town without everybody and their sister knowing about it. Not that that applies to me.

If you live in a city the size of Boise, or even Twin, I can't begin to explain to you what would happen if someone showed up at the Bliss Country Store with a big old black mole on her chin. If you live in a town the size of, say, Rupert, or Oakley, I don't have to. Cindy Crawford, my aunt Fanny.

It's the same in every small town, I reckon. Everybody's personal business is everybody else's public speculation. Who cares what happens in Zaire, really? If the lady three houses down has been entertaining a certain gentleman plumber three days in row, wink, wink, nudge, nudge, say no more, who wouldn't rather talk about that?

I certainly would. And I do.

While I am righteously appalled at the amount of erroneous, even absurd, gossip that disseminates about me, I am perfectly willing to share in any bald-faced lie about someone else. I mean, how bad can a person's plumbing really be? Heh, heh.

Another example might be, say, if a woman in town suddenly showed up with a black mole on her face, I would consider it my duty as a citizen of a small town to say to everyone I met for days afterward, "Hey, did you SEE that thing? What the hell IS that?"

Of course, gossip has been given a bad name by a few certain nosy priests and ministers and, I think, Jesus. But it has an upside. Bad marriages are not allowed to fester, bad parenting does not go unnoticed, bad teenagers are monitored like mice in a house full of cats.

And if it means everyone from Glenns Ferry to Wendell knows I had a pimple on my chin last week the size of a lap dog, so be it.

Now you know, too.

But don't spread it around.

Where good taste crawls off to die

By Suzanne Huxhold

CLOVER CREEK — OK, Steve Crump got to write about food for his fair column this year. I can accept that, but I'm pretty peeved.

Food is by far my favorite thing about the fair and everybody knows it. The fact that I didn't get to write about food just shows there are still, shall we say, certain prejudices in the newsroom. I'm every bit as well versed n the subject of food as Crump. I may not eat at McDonald's every day, but I've had my share. And Crump knows it.

This is chauvinism, plain and simple, and has nothing to do with the fact that I didn't write this column until the last possible minute. I'll just bide my time until I get an answer to my letter to my pal, Gloria. Then we'll see who gets to write about fair food next year.

In the meantime, I've had to come up with something else to write about the fair. (Not food, which is what I wanted to write about.)

So, here goes.

It is my feeling we need to institute some sort of dress code for the fair this year. And we need to do it before it is too late. Tube tops and knit shorts are going on sale even as we speak.

It's true, and you know it. I can see all the guilty looks on your faces as you read this. I can see you all mentally reaching into that little cubby in the way back of your closet where you keep your fair clothes — clothes you wouldn't be caught dead in at any other time of the year. I can see you all eyeing those crop tops and those shorts that leave the bottoms of your bottoms hanging out.

That's the problem, in a nutshell. I can see you all. Every square inch of you all. And I'm sick of it, I tell you. Just sick. I see more private parts on the midway on Saturday afternoon than I see all week in animal barns. The 4-H rams have more modesty than some of you. They're usually wearing T-shirts, at least.

It isn't just the women I'm complaining about here, though frankly it's got to the point where I have to buy those big clown sunglasses for my husband just so he doesn't get punched out from ogling all the women. Well, I mean, he does get punched out, but not by their husbands and usually not `til he gets home.

No, the men are far more disgusting. We're accustomed, in this society, to seeing women, um, less dressed. We get MTV on the satellite. We saw "Body Heat." We're used to that sort of thing. But you guys are truly an embarrassment.

Just what, for example, makes you think we want to see those huge expanses of white lard you've managed to stuff in your coveralls all year? Why do you think coveralls were invented in the first place? Your instincts during the remaining 11 months and three weeks of the year are correct. Go with them.

Furthermore, at least the women have the decency to tan the flesh they're exposing. You guys heave it all into those little polyester shorts you wore for football practice in 1975 and pair it with a tank top that shows more back hair than any human should have. I dread the annual parade of seared necks and chalk-colored legs. Put some pants on, for God's sake. Makes me want to puke. Really. I can hardly stuff down my elephant ears with all that running around.

So I'm proposing a logical solution. A dress code. With a few very simple rules:

1. If you're over 18 and over 160 pounds, don't wear short that expose your fanny. No cracks, no peaches, as my grandmother used to say. If you do and you have acne or hair on your butt, you'll go straight to jail.

2. People showing body parts they wouldn't show in church gotta go, too. So, too, do those exposing back hair for any reason (as if there could be any reason)..

3. If it isn't tanned, or at least fake-tanned, don't haul it out to the fair.

4. If you're a female, you must wear adequate support. I can't keep spending money on those big clown glasses.

I don't know what it is about the fair that brings out the worst in Magic Valley taste, but there must be something. A more tasteless display of carcasses I've never seen. (I'm sure I could relate it to the suspension of normal appetites and attitudes for one week in September and how that encompasses everything from the clothes we wear to the fact that we are willing to put in our mouths food cooked by great sweating women with pit hair and tattoos, but I'm not writing about food here, so I won't.)

Get dressed, Magic Valley. Put some decent clothes on. That means you, too, Steve Crump. I've seen you put away a Big Mac or two. I don't want to have to look at it walking down the midway in football shorts and a tank top.

CHAPTER 13
LIVESTOCK, OR YOUR SIX-STEP GUIDE TO BANCRUPTCY

We call 'em calf fries

By Suzanne Huxhold

CLOVER CREEK — I was kneeling down in front of a bull calf last Saturday, making him into a steer, when a burly fellow with breakfast still in his beard and a beer gut that hung over his 1972 high school rodeo belt buckle and threatened the integrity of his shirt buttons, came over to inspect my handiwork.

"Don't believe I've ever seen a woman doing that," he said. He said the word "woman" in that same lip-curled way my mother-in-law says "foreigners," like the very formation of it in his thoughts made him want to retch. "Who taught you how do that?"

"Well," I answered, plopping the extracted calf fries into a bucket and smearing pine tar onto the empty sac, "my daddy taught me this the summer I turned 5. He told me any little ol' gal who could rope and ride like I did oughta be able to do the same groundwork as her brothers."

"Five, huh?" he grunted.

I hopped up from the ground, picked up my bucket and pine tar and walked over to work on the next calf.

"Shore as shootin'," I drawled over my shoulder. "Why, I spent my sixth birthday a' larnin' how to cut a calf with nothin' but a big ol' sharp stick and a brandin' fire."

He went completely white at that. I went on.

"Yup, been doing this since..," I peered at his belt buckle, "well, since before you was champion goat tier at LaGrande High School."

"That's a horse. I was champion bronc rider, not a goat tier. Goat tying's fer girls," he sputtered.

"Sorry," I said, smiling my biggest, most feminine smile. "I thought it was a goat."

"Well, whatever yore daddy said, my opinion is having a woman cut calves is breaking with tradition," he said. "There is things for women to do at brandings at things for men to do."

"And just what is you are doing at this particular branding?" I asked, looking at the beer in his hand.

"Uh, well, I came to rope but it looks like there's plenty a ropers, so I'll just wait my turn."

"Good," I said. "In the meantime, carry this bucket around for me, will ya, honey? It's slowing me down."

He sheepishly trotted after me for half an hour, until the dinner bell rang. When we went in to eat, he wouldn't even sit in the same room with me. Which was just as well, because my husband, who had listened to the whole exchange, sat down next to me.

"What kind of lies were you telling that poor guy out there?" he asked, laughing. "Your father digs septic tanks in Kingman, Arizona. I bet he's never even seen the underside of a calf, much less taught you what to do with one."

"That's true," I said, shoveling a forkful of mashed potatoes in my mouth.

"And you just learned to castrate two Saturdays ago, and you certainly do not know how to rope a calf," he continued.

"That's true, too," I said.

"So why were you saying all that about a sharp stick and a branding fire?" he asked.

"Because you guys all act like roping and castrating and ear-marking are some kind of holy secret, passed down in your Y-chromosome," I said, spearing his dinner roll with my knife and sliding it onto my plate. "I'm sick of all this tradition. I want to be where the action is."

"You always are where the action is," he said. "I can't remember once ever hearing anybody tell you to stay in the house."

"Sure, but all I ever get to do is run for beer or fill the vaccinating needles," I said. "I want to get smoke in my hair and dirt in my eyes and flank a fall bull just to see how strong I am. It makes me feel alive."

"You mean it makes you feel like a man," he said, his eyes twinkling.

I smiled back at him.

The big burly guy with breakfast in his beard came by just then.

"Hey, pudding, grab me a beer while you're up, will ya," I said.

He did, and brought my husband one, too. But I noticed later he never did come back to the branding fire. Probably stayed inside to help with the dishes.

Too bad, really. He was pretty handy with that bucket.

Cow emergencies

By Suzanne Huxhold

CLOVER CREEK — We were watching one of those new hospital shows the other night, I can't remember which one, "Chicago E.R." or something like that, when one of the TV doctors shouted to one of the TV nurses, "We've got an obstruction in the lower intestine! I'm gonna open her up! Get me a big old knife! Stat!"

"Eew," my husband said, wrinkling his nose. "I could never be a doctor. It's too disgusting."

I stared at him for a minute, sure I didn't hear correctly.

"What do you mean?"

He picked up the remote and began scanning the airwaves for something less repulsive.

"I mean, I could never do that kind of stuff. Opening people up, looking around inside their bodies. It's gross. I can't see how doctors do it."

"It's their job, for one thing."

He settled on the Cartoon Network. They were featuring a "Scooby Doo" retrospective.

"Well, I'd look for another line of work, if I were them."

"Like...rancher, say?"

He scowled at me. "You're about to make a point here, aren't you?"

"I'm just saying I've seen you do things in the last ten years that would make Marcus Welby vomit."

"I have not!," he said, his eyes widening.

"Oh, please. I once watched you heat your hay knife in a branding fire and dig out the eye socket of a cancer-eyed cow."

"That was ONE time," he protested.

"Huh. I've seen you go into the back end of a cow to your shoulder and dig out handfuls of cottage cheese-looking stuff so brown and stinking I was sick for a week just from the thought..."

"You have to do that if they don't slough their afterbirth," he interrupted sensibly.

"...and then go back in with pills the size of your fist and then have to hold your hand under cold water for five minutes because an infection had baked her insides so hot it burned your skin..."

"Yes, well, that stuff almost never happens."

"Ha! Remember two years ago when that heifer got bit by a rattlesnake and you didn't see her until a week later..."

He put up a hand to stop me. "O.k. I remember."

"... and by that time she had that cyst on her jaw that looked like a second head and you had to slice it open and drain it and the pus was as green and viscous as lime Jell-O and the smell made that one cowboy of Jim's pass right out there in the corral..."

He smiled reminiscently. "Yeah, that was disgusting, wasn't it?"

"...and how about the time you pulled that calf that had been dead for a couple days inside its mother and the hoofs came off in your hands..."

He eyed me. "Will you be wrapping this up soon? Scooby and Shaggy have almost figured out how the crooks got those diamonds in the dolphin tank."

"...and how about in spring when you have to treat everything that goes through the chute for lice and how about SCOURS, for crying out loud? How many doctors in "Chicago E.R." meet up with scours, do you think? And what about pink eye? Or mastitis? Or the time you found that calf who'd got wire-wrapped and her leg came off at the knee and you had to hold her on your lap and pick the maggots out before you could clean the stump? Or the cow that died of an aneurism with a live baby inside her and you couldn't wait for the vet so you performed a C-section right out in the field and saved the calf...?"

"Are you through?"

"Pretty much." I smiled. "I do tend to go on, don't I?"

"You do. I guess I never thought of any of the things I do as disgusting. In fact, I like it most of the time. There's nothing more interesting than doctoring a cow and or watching a calf get well because of something you've done."

"I expect doctors feel the same way."

"I expect they do." He watched "Scooby Doo" for another minute, then turned to look at me. "But I don't expect their nurses have ever described afterbirth as cottage cheese."

"I bet they do." I snagged the remote and turned the channel back to the doctor show. "Just not on TV."

You want farming?

By Suzanne Huxhold

CLOVER CREEK — It's been a tough farming summer around our place. The equipment, on its last legs, breaks down with uncanny regularity, and the weather has been even less cooperative. Wet when it should be dry, dry when we need it wet, and dead cold in the middle of June, for crying out loud. The men are so far behind even they can hardly believe it, and these are men who are no strangers to being WAY behind. The first cutting is just now in the stackyard and already some of the second cutting is days past perfect baling.

So it was good opportunity to get my foot in the door. I've always been intrigued by farming. There's a mystical quality to it. Man and nature, dirt and sweat, manure and swearing. I've been wanting to get in on it.

But for 10 years, Chuck and Scott have been putting me off, citing my allergies, my child, my utter incapacity for driving any vehicle under 65 mph - "That'd be mighty tough on the tractor across those corrugates," they sensibly point out - as reasons why I couldn't. But I've worn them down. Last night my husband said I could spend the morning mowing the ditches; the lowliest of farm chores to be sure, but I'm not proud. I jumped at the chance. Then the alarm on his watch went off at 6 a.m. I turned over and stayed asleep, as I always do, but he shook me.

"WHAT?!"

"It's time to get up."

"Yeah, right."

"No, it really is. Get up. You want farming, this is farming."

I tried to bat my eyes at him. "Couldn't I do that silly old mowing this afternoon, hmmm?"

Unmoved, he said, "We need the tractor this afternoon to pick up pipe. Now get out of bed."

I got up, RELUCTANTLY, and dragged on a pair of shorts.

"Jeans," he barked.

"But it's already so ho-o-o-o-t," I whined.

"And boots."

"Boots? My feet will sweat."

"You want farming, darling," he said, patting me, "this is farming."

Sweltering and un-breakfasted, I met him in the driveway 10 minutes later. He was wheeling the compressor out of the shop. I looked at the John Deere. One of the new tires, the one we had to order from Michigan and wait three weeks for and pay for in cash because 35 years ago we bought the one John Deere in the country with 36-inch rims, was flat.

"They just put that on yesterday," I shrieked.

Scott didn't even blink. "You want farming," he said evenly, "this is farming."

"What about breakfast? What I want now is breakfast."

"Later." We pumped up the tire and hooked up the mower. By then, the tire was flat again.

"That SEEMED like a waste of time, though I could be wrong," I muttered.

"You are wrong." We went to the house, for breakfast, I hoped.

But he only called the tire store, headed back outside and jumped in the pickup. "They won't be here 'till 8," he said. "You can help me start water."

Well, now he was talking! I forgot about my stomach and the heat and climbed in beside him. THIS would be real farming at last!

Then we drove to the tack shed, gabbed a pitchfork, drove to the shop, grabbed a shovel, drove to the hay yard, grabbed a plastic dam, then drove to the barn, grabbed a different shovel (!), then drove to the field. Well, um, that was tedious, I thought; but I consoled myself with the thought that soon I would be farming!

Wrong again. I drove the pickup over 40 acres of corrugates I defy even a WonderBra to withstand while he forked broken bales into the back of the truck, then I watched him set dams for a half hour. God, I was so bored.

"Um, honey," I ventured, "when will we start the farming?"

"You want farming, this is farming."

Uh huh. When we got back to the house, the tire guy told us he had to back to town to get a part and the tractor would be ready by Thursday. I wanted to scream in frustration. Scott merely shifted gears.

"Well, that'll give us time to get that south 80 swathed, then. And I've got to do a little work on the wheel line pump.

And..."

"WAIT!," I cried. "What about me? What about mowing? What about FARMING?

"You want farming, this is.."

He didn't finish. The tire guy said later I hit pretty good, for a girl.

I changed my mind. I don't want farming. I don't want it at all.

Like manna from hell

By Suzanne Huxhold

CLOVER CREEK — It's like living in the Bible around here. Not the good parts, though. Not the love thy neighbor and forgive thine enemies parts.

Nope, we're living in the OLD Testament. We're living the scourge and pestilence sections of the Bible. We're talking Ecclesiastes here. We've got bugs raining down from heaven, mammal-eating frogs, crickets the size of your fist. I don't know why. We haven't been worshipping golden calves or anything.

Still, we've got everything from army worms and snakes and killer toads as big as dogs to gnats and flies and moths. We've even taken to eating the horseflies, they're so big and juicy. And the ticks? Well, you can fill a false idol with the ticks we've taken off our kids around here this summer.

But nothing compares to the mice. Except maybe the rats.

We've been combatting our furry Philistines in a variety of ways. My husband, as I told you last week, prefers the aggressive approach. He likes traps. He likes everything about traps. He likes setting them in strategic places, the sound they make when they snap, the little squeaking and flopping noises from under the sink when the mouse is really big or caught by just his tail and so doesn't die for a hour or so...everything. He even likes shopping for traps.

"Aaah, the Mouse-Masher 3000," he'll say, his eyes glistening, his tongue licking the corners of his mouth. "I read about this little baby in Mouse and Ammo. Oooh, lets get some of these."

I, on the other hand, prefer something less bloody. I loved the idea of poison, until I realized I'd have to live in the camper until winter. Have you ever had 1,100 mice die in your walls in 90-degree heat? Yeah, well, talk about Old Testament suffering. MOSES didn't suffer like this.

I thought the perfect solution to the mouse infestation was to allow the cats to stay in all night. Our cats are brilliant mousers. During the fall, they often abandon cat food for weeks in favor of eating the mice in the haystacks. So I let the three cats in every night for a week. Big mistake. Huge.

The first night, the first HOUR they were inside, they caught three mice; one each. But they didn't kill and eat the mice as they do in the haystack, leaving nothing but a crunchy little head and a half-chewed tail. No, they played with them. For HOURS. We just laid in our bed and listened to them scurrying under the couch, through the television cabinet, across the piano, over the ironing board and up and down the ficus tree in the living room.

Oh well, I reasoned, at least they're catching them.

The next night, I foolishly risked going to the bathroom in the middle of the night while three more little mice were meeting their doom at the paws of my playful assassins. When I flicked on the light in the living room, however, I found the tremendous noise this night was not made by a cat and mouse game, but by a cat and RAT game. One cat, John-Darling, had a gigantic pack rat by the tail and was alternately catching and releasing it.

"Scott," I croaked, gasping for air. "Scott, Scott, Scott, SCOTT, SCOTTTTTTT!"

He came running out in his underwear and tried to get the live rat, which was the size of a small car, away from John-Darling, who was appalled enough at the prospect to take his rat immediately to the safest spot he could find, UNDER MY SON'S BED, and hide it there. Scott finally got the cat and rat by their respective tails, and then beat the rat to death with Joseph's hockey stick. We all had happy dreams THAT night, I can tell you.

I let the cats in the third night for one more try. I heard the occasional scuffle during the night, but I was getting used to it. I was more bothered by the heat, frankly, which was why I was lying on top of the covers when John-Darling and his brother, Copper, came screaming across my legs, chasing, yes folks, a mouse.

In my sleepy stupor, I couldn't really figure out why they were on the bed until I realized not ONLY had they brought their live rodent onto my bed, but that it had CRAWLED up my bare leg to hide in a strategic spot under my nightgown.

If you were in Pocatello that night and heard me screaming, I apologize. You don't know how loud you can scream until you have a mouse hiding in your nightgown and two cats trying to find it.

We've gone back to traps, obviously. And until I get a sign from God- a burning bush, a parting of the sea, you know, the usual- we're sticking with that program.

CHAPTER 14
TOTO, I DON'T THINK WE'RE IN L.A. ANYMORE

You're not from around here, are ya?

By Steve Crump

An extraterrestrial – little green fella with four eyes – landed in the backyard the other day, and we got to talking.

"We heard on the planet Zarmak that you Idaho humanoids are especially gullible," he said, scratching his third ear. "So my name is Zellnak and I'm from France."

I was proud to shake his antenna. What with all the talk in these parts about Hale-Bopp-riders, underground aliens and UFO abductions, I was afraid he was from California.

There are certain similarities, you know.

Both expatriate Californians and extraterrestrials drive Saturns, wear Planet Hollywood T-shirts and occasionally frequent the ozone. And I know they're here because I followed a car down North Blue last week that had California tags framed by a license-plate holder that read, "I want to have Spock's baby."

Do you really think it's a coincidence that the Jefferson Starship is getting back together for a reunion tour after all these years?

Could it be an accident that every single motion picture and TV show that trumpets the arrival of visitors from the beyond, from "Independence Day" to "Millennium," is made in the Golden State, or that Jerry Brown hasn't been seen in public for weeks?

Perhaps the Guru of Zen and the Art of Direct-Mail Fundraising really has gotten in touch with a higher power.

Jerry Garcia, rest his soul, got it right when he observed that California is on a different plane. I don't think he had an Airbus in mind.

Alien abductions? Please. They've all come here and bought ranchettes.

Anyway, I invited Zellnak into the house for some Napa Valley zinfandel and goat-cheese brie, and we talked about beach-front property and killer commutes. He does the 1.2 million light-years from Andromeda Prime to Malibu five days a week, and the traffic is starting to wear on him.

That's why he's shopping for a nice small-town Idaho Main Street colonial with room to add a docking bay for the Mothership.

Far out, dude. What kind of mileage you get with that Beamer?

A few basic life rules for Californians

By Steve Crump

Back in the days before Californians brought us colored beads, trinkets and Mobilink, we Idahoans dwelt in darkness.

Thousands of us, in fact, went to bed before 10 o'clock. Others drove half-ton Ford pickups repainted in primer-red, and we parked them right on Main Street.

Still others had never been to a health club, tasted brie, guzzled weak beer at a microbrewery, slept on a futon, been thanked for not smoking or eaten tacos with anything but mayonnaise-and-ketchup sauce.

Rejoice, brothers and sisters – the scales have fallen from our eyes.

The first wave of California refugees that grew disgusted with our loutish ways has returned to the Golden State and chronicled their experiences to their fellow lotus-eaters.

And let me assure you, friends, theirs are tales that will raise the hair on the back of your red necks.

"Once the beauty wears off," Kathy Smith, a once and future Orange County resident late of Coeur d'Alene, told the Los Angeles Times recently, "there's not much to like about this place."

"I was lonely," Barbara Lund said. "Men were afraid of me because I was a city girl. They didn't have a clue. One wanted to race me to the top of a hill to see if I could keep up, gauge what kind of woman I was. I thought he wanted to inspect my teeth too."

"You show me someone who doesn't complain about the smoke and I'll show you an idiot," John Mann said. "But these farmers are given free rein — just because this is the way they've always done things.

"I just don't get it. This would never happen in California."

Mr. Mann, God bless him, has had an epiphany.

We troglodytic knuckle-draggers up here in God's own country really don't *want* things to happen the way they do in California. Truly. That's why we're still wandering here in the wilderness, miles from a nearest frozen-yogurt stand.

We *like* the way we live, thank you so much, and we want no part of the California cocktail. Shaken but never stirred, it consists of equal parts greed, paranoia, angst and anarchy, and comes highly recommended by the likes of Charles Manson, Richard Nixon and Snoop Doggy-Dog.

So in the interest of Western comity, here's a list of rules I propose tacking on the bottom of the "Idaho Is Too Great to Litter" sign just north of Jackpot for the benefit of our new neighbors:

1. Above all, we Idahoans loath pretense. It will shock you, appall you, and shake you to your California soul to discover how little we care what toys you've accumulated, and we're genuinely unsympathetic to the fact that you can't find snow tires for your Beamer here.

2. Don't try so hard to improve us. We're far from perfect and prone to folly, but we're comfortable with who we are. Can you say as much?

3. There's a difference between being a hayseed and being a bigot. Many of us who happily admit to the former want nothing to do with the latter, nor do we appreciate the easy characterizations of strangers from a land where hate crimes happen every day.

4. No, there is no Top 40 radio station in this market. Yes, there are six country stations in this market. No, that fact doesn't bother us.

5. If you don't know what you're talking about, don't.

6. It's of surpassingly little interest to us how the Dodgers, Angels, Lakers, Clippers, Raiders, Rams and Kings did last night.

7. Don't confuse quiet with boredom. With ex-Californians, that's a dead giveaway.

8. Despite what you may hear, Idahoans are *not* innately hostile to Californians. What we're hostile to is what happened to your old neighborhood while you were living there.

9. Motion does not necessarily equal progress, much less happiness. Many of us Idahoans are perfectly content just to *be.* If you are not, have the goodness not to bellyache about it.

10. Don't keep reminding us that we're laid back or that life is slower here. Why do you think we live here?

11. There are 1.2 million Idahoans, and you can count on the fingers on one hand those who want to hear how much you got for your house in the San Fernando Valley.

12. The median income in Twin Falls County is $26,821. The median income in Idaho as a whole is $29,620. If that is not adequate to your needs, may we suggest that you go back to your four-hour commute to Westlake and leave us to our 10-minute commute to a trout stream?

13. Being an immigrant yourself in these parts, please don't let us hear you complain how the foreigners ruined California.

14. We Idahoans drink bottled beer. We don't drink bottled water. Why on earth have you migrated to the Rocky Mountains if you're afraid to drink what comes out of the tap?

15. Stop bragging about the good life you left back home, lest you be invited to resume it.

Say, that buffalo looked at you funny.

By Steve Crump

Idaho is getting way too dull. We got Phil Batt, $8 spuds and the world snaffle-bit championships.

We need some adventure. We need an urban legend.

Urban legends are phenomena – long on imagination and slim of substance – that city folks invent to distract themselves from the fact that they're scared to go outside their $1,200-a-month apartments and that they have no lives.

Batman was an urban legend. So were the Green Hornet and Superman – avengers of the milquetoast, vindicators of the bemused.

So were Shaft and Charles ("Death Wish") Bronson, for that matter. But the best urban legends have beady eyes and real big teeth.

Every city dweller has heard stories about three-foot-long rats and 12-foot gators who've lived in the sewers since they were flushed down the toilet by sadistic urchins and are now lurking in the shadows to swallow the selfsame brats whole now that they're taxpayers.

The reigning champion urban legend is a critter called the goatsucker, a flying, kangaroo-like beast with bulging, red eyes that's haunting Mexico, the Caribbean and parts of Europe and the United States.

Strictly speaking, there's no such thing as a goatsucker, or chupacabra, as it's called in Spanish – a fact that most folks in chupacabra country choose to ignore.

Goatsuckers are supposed to feed on the blood of goats and other small farm animals, and although no human has ever reported being bitten, well, there's always a first time.

From San Francisco to Moscow, mothers are keeping their children indoors, and according to Knight-Ridder Newspapers, the mayor of the Puerto Rican city of Canovanas, where the legend began, is convinced that the goatsucker is an alien.

I don't see the resemblance to Sigourney Weaver myself, but the point is that goatsuckers are real-for-real, scare-the-bejeebers-out-of-grandma, what's-that-scratching-noise-at-the-window bogymen.

Think of the fun we could have with the Californians: Let's invent our own urban legend.

Giant jackalopes would be good candidates. The half-jackrabbits, half-antelopes have been on postcards for years, and maybe one visitor in three to our little corner of the weed patch actually expects to encounter one astride the freeway, squaring off with oncoming tractor-trailers.

Wooly mammoths would be winners too. They look like unshaven elephants, which would attract a lot of Republicans. And a wooly mammoth footprint is an instant tourist attraction, provided that it doesn't enclose a flattened Volkswagen.

Giant sloths are a possibility – nine-foot-tall, nice, fuzzy leaf-snippers, sort of Barney in earth tones – but we're a little thin on trees around here and I don't want to be the one to break the news to the Idaho Cattle Association that our state bugaboo is a vegetarian.

So let's stick to basics and go with the beefalo – marauding, bad-to-the-bone furry mutant bovines in an endless search for the descendants of Buffalo Bill Cody.

We could put up billboards with a picture of empty cowboy boots over the caption, "His Justens are still here, but where's little Biff Cody?"

The implication, of course, would be that Biff ran into some beefalo who knew more about his genealogy than Biff did.

Ponder the possibilities: By now, Bill Cody has – what? – maybe 5,000 blood descendants, many of whom have no idea that they're kin. Anywhere they wander in Idaho, they'd never be more than a few miles from a cow.

Toto, I don't think we're in L.A. anymore

And any of *one* those cows could be closet bison.
I, for one, want it to be known that none of my forbears here in the West shot anything more noble than a Spam.
There's no such thing as a Spamalope, is there?

Chapter 15
Mormons and gentiles

Rubbin' elbows with Cousin Brother Pugmire

By Steve Crump

There he stands, this fat, balding guy in Bermuda shorts and a Hawaiian shirt that can't quite make ends meet.

He stuffs what's left of a hot dog into this mouth, chases it with a gulp of Old Milwaukee, wipes the mustard on his shirt and extends his right hand.

"Hi, I'm Daryll. Who the hell are you?"

Welcome to the family reunion.

Thirty years ago, Cousin Daryll was a skinny kid with a shock of red hair who was the terror of events like this. With Daryll nearby, bullfrogs somehow got into the potato salad, the lemonade tasted of Uncle Ed's sloe gin, salt shaker lids were loosened and condiments were laced with Tabasco sauce.

Daryll was widely considered to be subversive enough to be banished to Uncle Ed's camper, except for his artful ability to convince his mother, Dora, that he'd been framed.

"How *could* you accuse Daryll of such a thing?" she'd shriek, her voice quavering with righteous indignation and Darryl clutched close to her bosom, tears streaming down his face. "I teach *my* boy how to behave."

Now, a generation and a half later, Daryll's eyes brighten in recognition.

"Stevie!" he bellows. "Cousin Stevie! I haven't seen you since that time you set fire to the outhouse at Indian Springs during Grandma Phyllis' 75th birthday picnic."

It's ever thus at my family's reunions, periodic gatherings that are just seldom enough to induce widespread guilt about family ties neglected and just frequent enough to make everyone wonder why we don't do this less often.

Like a lot of other families that have lived in the Intermountain West for generations, mine has Mormon and non-Mormon branches, which means that cousins can live literally next door to each other and never know they're related until they're standing in the same line for barbecued chicken.

"You fixed my brakes last month, didn't you?"

"I don't think so. I live in Salt Lake."

"So do I."

"In that case, they still don't work."

"I'm glad to hear that. Your check bounced."

"Serves you right, taking a check from somebody you don't know."

"What? You're my cousin."

"Really? Pleased to meet you."

Back in the days when my grandparents were still alive, this was an easier process. They remembered Aunt Edna, Cousin Farnsworth and Nephew Enos, the black sheep of the family who became a lawyer. Now it's a matter of wading into a gaggle of strangers and convincing yourself that you're really related to these people.

"Hi! My name's Steve and I'm from the Idaho branch of the family and you don't look like anybody I've ever met."

"Hi! My name's Bob and I'm from the Utah branch of the family and you look just like Great-Uncle Ernie before his tragic accident."

"Tragic accident?"

"He was a blacksmith and got kicked right between the eyes by a Clydesdale. Ended his life as a clown, doing magic tricks at kids' birthday parties."

"Sounds harmless."

"He wasn't invited."

Actually, it doesn't take that long to break the ice at a reunion. People are people and kinfolks are kinfolks, and once you get enough memories, cole slaw and Aunt Anita's rum cookies into you, it's not that hard to forge a family bond.

I think the most fun I've ever had was at a reunion a few years back in a park in Pocatello. We arrived as strangers and, by the time we were ready to leave, had become friends and well as relations.

It was only when we exchanged phone numbers that we discovered we weren't. The Crump family reunion was the *next* Saturday.

Still get Christmas cards from the Pugmires, though.

What do you mean, `Hold the guppies?'

By Steve Crump

For those of you who think that Idaho cuisine is about as exciting as Friday night at the Feed 'n Ice, have I got some salads for you.

Announcing the winners of the First Annual Don't Ask Me All-Idaho Gelatin Surprise Recipe Contest, dedicated to three generations of pretty darn mediocre Mormon cookery:

Grand prize winner (hands down, I might add): Cherie Frisby, Twin Falls.

GOOD GOD GUPPIES JELL-0

3 packages lime Jell-O
3 packages strawberry/banana Jell-O
1 pound ground salmon
1 can of creamed spinach soup
1 dozen live guppies
4 cups of water

Mix Jell-O according to instructions on package. Add salmon and soup. Refrigerate until Jell-O has congealed. Add 4 cups of water on top of Jell-O and add live guppies. Let swim for five minutes before serving.

(Editor's note: You can substitute gummy bears for the guppies and tuna for the salmon. Serves 12, but that'll be more than enough. Trust me on this.)

Cherie wins a certificate and a case of Jell-O of our choice. I'm thinking peach melba.

Second prize: Kathie Dunlap, Twin Falls.

BARBECUE SALAD

1 package Jell-0, any flavor
1 cup boiling water
1 8-ounce can of tomato sauce
1 1/2 tablespoons of vinegar
1/2 teaspoon of salt (optional)
Dash of pepper
Additional seasonings (*Editor's note: Kathie is talking Tabasco sauce or prepared horseradish here, but I wouldn't do that if you have a gas range).*

Dissolve Jell-O in boiling water. Add remaining ingredients. Pour into individual molds and 1 quart mold. Chill until firm. Unmold on crisp greens. Makes about 2 1/2 cups.

Possible Barbecue Salad additions: 1/4 cup of mayonnaise, 1 cup of cottage cheese, 1 cup of diced celery, 1/4 cup of thinly sliced stuffed olives, 1 cup diced raw or cooked vegetables, 1 cup of diced cooked shrimp, 1 can of crab meat, 1 can of tuna drained.

Mercy.

First prize, lime Jell-O Division: David Bell of Ketchum.

IDAHO POTATOES AND GREEN JELL-O

"One day there was going to be a potluck (at the Ketchum Fire Department) and I wasn't a great cook. However, I was tainted by living in the prestigious and not real world of the Wood River Valley. I knew that Jell-O was not food for 'real' people. It was for hospital patients and others less fortunate that couldn't get to the Western or Louie's. The wheels were turning, and it hit me like the flu:

"Idaho potatoes and green Jell-O.

"So I carefully blended raw strips of potatoes with green Jell-O (I don't remember the flavor, I don't think it matters, as long as it's green but since there is a separate prize for lime flavor, I'll say lime). I didn't think anyone would even put it on their plates. I was wrong. These people not only put it on their plates, but put it in their mouths. I heard them saying, 'The potatoes don't seem cooked,' and, 'Who made the Jell-O?' I was happy. At the next year's potluck, I put Styrofoam peanuts with ketchup under the heading of shrimp cocktail. People wouldn't try anything I brought after that.

"Go figure."

Congratulations to the winners, and the best of all possible luck to the people they cook for.

Now, would you stop talking to your Cousin Dennis and pass the lemon buns?

It's nearly Lent: Wipe that grin off your face

By Steve Crump

I'm sort of a theological thrill-seeker.

My dad was raised as a Mormon. My mother was Baptist. My best friend is Jewish. My first college roommate was a Muslim. The second was a Hindu. I married a Presbyterian, and we sent our kids to Immanuel Lutheran.

But of all the doctrine and dogma clanging around my cranium, none has more resonance than the eight years I spent in Catholic school.

Especially this time of year. When you've been to Catholic school, Ash Wednesday – that's the day after Mardi Gras for you secular humanists – never passes without a good deal of reflection.

OK, I mean a good deal of guilt.

Ash Wednesday starts Lent, which is 40 days of applied soul-searching. Whether God or the Pope had it in mind or not, Catholic kids – and non-Catholic kids who go to school with them – learn to behave for six weeks as if they were sitting in the principal's office.

By tradition, Lent requires some sacrifice – no sweets, no TV, no fighting with your siblings. In short, it sort of drains all the joy right out of life.

I always thought of it as never-ending family reunion at your Aunt Bertha's, during which you had to sit there, in white shirt and bow tie, on a metal folding chair, forbidden to squirm.

A lotta kids just couldn't take the pressure. I've seen Jujubee-binging before the first of March and hooky-bobbing in broad daylight. Jerry Bryant and Mark Lewis were even caught after school one day dealing 21 with holy cards.

As a concept, Lent – which culminates at Good Friday and Easter – is a fine idea, but my friends and I never seemed to get the point.

It was just six more weeks we had to hide from Sister Marietta.

Sister Marietta was the top kick at our school, and her approach to the righteousness of the young souls in her care was proactive.

Spare the rod, her philosophy went, and the kid's headed for 10 to 20 in San Quentin.

So she sacrificed many sticks – heavy oak pointers that she must have ordered in bulk from the Office of the Inquisition. Smirk at the wrong time, pass a note in class, or turn to inquire of a classmate the square root of 42, and you'd hear the terrible *Twaaaack!* of Sister Marietta's terrible swift sword across your desk – or occasionally across the knuckles of any hand sitting thereupon.

You weren't merely busted. You just *knew* you were going to do hard time in someplace sulfurous.

For although Sister Marietta never shied from leading an urchin by the ear down the corridor and toward retribution, her greatest impact was psychological.

You could tell that in morning prayers, which we'd all stand next to our desks and recite. When we got to the part of the Act of Contrition that said, "... and I confess all my sins ..."

"Quiet!" she would say in a loud whisper, then continue.

" ... because I fear the loss of heaven and the pains of hell ..."

It was plain enough. The woman had a direct line to St. Peter, and she was taking names.

The worst offenses, in her book, came during Lent, when introspection and self-control were deemed to be the hallmarks of a pure heart.

That's a stretch for any fourth-grader.

One particular Lent, the boys in our class took to making rude animal noises whenever quiet would descend over the classroom for more than a few seconds. Sister Marietta welcomed this like Spam at Friday's lunch.

"If I hear *one more sound* before Lent is over, I can guarantee you people that you'll be spending the 21st century scrubbing the basement of Purgatory," she said. "*On your knees.*"

The basement of Purgatory was clearly not that far from hell, and Sister clearly knew the manager.

No more animal noises were heard in that classroom until "Away in the Manger" the following Advent.

Even today, I have trouble getting through Ash Wednesday without thoughts of a vast marble floor being scrubbed by a pitiful urchin with a toothbrush and a bucket of lye.

Me, for example.

It's so unfair. I'm not even Catholic. Would you mind rapping my knuckles with that ruler?

And now... a family disclaimer...

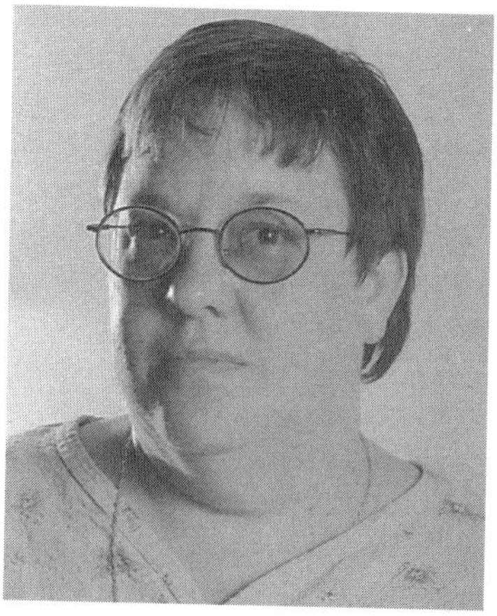

By Connie Crump

Over the years, many people have asked me where Steve gets his ideas for his column. Many times, the source is very obvious — my family, his battles as a homeowner. Others have come from the clear blue sky.

From reading his columns, you might think that Steve has a very large extended family. Not so, I'm sorry to say. From his vivid imagination (and hallucinations) he has created family members that all of us can claim.

My children, Michael and Eric, and I thank all of Steve's fans for the many gift selections he receives. Please, no more black velvet paintings of Elvis.

By Scott Huxhold

Living with Suzi for 10 years has taught me to laugh at just about everything, including myself. Her column has taught everyone else to laugh at me, too. But I'm in good company. She makes fun of other people almost as often as she makes fun of me.

I'd really like to say to enjoy this book, but to take it with a grain of salt, because it's not all true. But I can't say that, unfortunately. It is all true, although some stuff is exaggerated. But not much.